PRINCIPLES OF WATER QUALITY

Water Resources and Water Quality Management

An International Series of Books

EDITOR

Hillel I. Shuval

Environmental Health Laboratory
Hebrew University—Hadassah Medical School
Jerusalem, Israel

Thomas D. Waite, Principles of Water Quality, 1984

PRINCIPLES OF WATER QUALITY

Thomas D. Waite

Department of Civil and Architectural Engineering
University of Miami
Coral Gables, Florida

With Chapters by
J. E. Quon†
Department of Civil Engineering
Northwestern University
Evanston, Illinois

Neil J. Freeman
Department of Civil and Architectural Engineering
University of Miami
Coral Gables, Florida

1984

ACADEMIC PRESS, INC.

(*Harcourt Brace Jovanovich, Publishers*)

Orlando   San Diego   New York   London
Toronto   Montreal   Sydney   Tokyo

ACADEMIC PRESS, INC.
Orlando, Florida 32887

United Kingdom Edition published by
ACADEMIC PRESS, INC. (LONDON) LTD.
24/28 Oval Road, London NW1 7DX

Library of Congress Cataloging in Publication Data

Waite, Thomas D.
   Principles of water quality.

   (Water resources and water quality management)
   Includes index.
   1. Water quality. I. Title. II. Series.
TD370.W34   1983   628.1'62   83-3899
ISBN 0-12-730860-1

PRINTED IN THE UNITED STATES OF AMERICA

84 85 86 87   9 8 7 6 5 4 3 2 1

This book is dedicated to the two most courageous individuals I have ever known. The first is Professor Jimmie Quon, whom I was privileged to know at Northwestern University, but for only three years. The second is my little nephew Scott Shelly. Both fought a losing battle against cancer. Their tenacity and unrelenting optimism will always remain an inspiration to me.

Contents

4. Refractory Organic Compounds and Water Quality

5. Nutrients, Productivity, and Eutrophication

6. Microorganisms and Water Quality

7. Thermal Effects on Water Quality

8. Potential Impact of Air Contaminants on Water Quality (by J. E. Quon)

9. Water Quality Modeling (by Neil J. Freeman)

10. Water Quality Standards and Management Approaches

Preface

"Principles of Water Quality" has been created for two specific purposes. First, it attempts to introduce engineering and science students to the fundamental environmental processes that regulate the movement of materials in natural systems. Because all of the treatment schemes employed to purify water and wastewater are engineering models of natural phenomena, it is important for students to understand these underlying processes. With this in mind, the text details the chemical and microbiological processes that are operative on organic and inorganic constituents in water. Emphasis is placed on those constituents that have become water quality problems, e.g., toxic metals, inorganic nutrients, and various organics. In each case natural systems are described, along with the normal process kinetics involved. Human input of various contaminants is then described, including possible health effects and impacts on natural systems. Thus, the student will be able to understand which contaminants are true water pollutants and which do not pose severe threats. This information will then allow the student to understand treatment approaches more clearly.

The second purpose of this text is to introduce engineering approaches for water quality improvement other than sewage treatment. All the processes described are quantified as much as possible, so that predictions regarding water quality can be made. Microbial and chemical kinetics are described for many phenomena, and the empirical data required for the determinations are also presented. Actual projects with respect to water quality restoration are described and quantified so that more rational approaches to these problems can be made. The information supplied will provide the student with methods for treating water systems of low water quality, including lakes, ponds, and streams.

This text is intended for advanced undergraduates or graduate students in environmental engineering and science and in health-related disciplines. A familiarity with physical chemistry and microbiology is presupposed, but students should be able to understand the concepts even if they have not had formal courses in microbiology and aqueous chemistry. This text can be used ahead of the classical sanitary engineering courses or for a "water pollution" course at the end of the normal course sequence.

This text represents a culmination of efforts by many individuals. I am especially appreciative of the chapters contributed by Dr. J. E. Quon and by Dr. Neil J. Freeman. In addition, I am grateful to Marsha Gilbert, Ram Tewari, and Kimberly Gray for their efforts and suggestions.

Chapter 1

Introduction to
Water Quality Concepts

I. THE NEED FOR WATER QUALITY STUDIES

Understanding the water quality of natural aqueous systems requires knowledge of the disciplines of chemistry, microbiology, ecology, and engineering and how they interact. Although the areas of water and wastewater treatment require the application of biological and chemical phenomena, neither of these areas is as heavily dependent on application as is the study of water quality. A great deal of literature has appeared in the past 30 years that shows how natural processes can be applied to treat wastes of various types, and the sophistication of treatment techniques has increased only with an understanding of these natural processes. Quantification of many of the specific processes required in waste treatment has evolved rapidly, owing to the ease of containment of these processes in reactor systems.

The quantification of similar processes in natural systems, however, has remained relatively unexplored. The complexity of most natural systems, for example, streams and lakes, has precluded the rigorous quantification of many common natural processes. In addition, so many parameters affect

1

natural systems that it is difficult to isolate the individual units of the process. Therefore, a scientific or engineering approach to problems of water quality is still unrefined. As an example, it is still almost impossible to predict the environmental impact of discharges of certain contaminants into aqueous systems. This is owing in part to the fact that we do not fully understand the natural processes responsible for the movement of the various contaminants. It is imperative that the specifics of these processes be understood, for they determine the degree of wastewater treatment required before release. Because each level of waste treatment costs increasingly larger amounts of money, a quantification of the resultant environmental damage caused by contaminants is important. Indeed, many environmental contaminants have very low threshold levels and should never be discharged into aqueous systems.

Perhaps the most important reason for studying water quality is the increasing worldwide demand for potable water. Many civilizations in the past failed because of a lack of fresh water. Recent evidence indicates that certain parts of the United States will run short of fresh water in the very near future. It is now apparent that we should protect all water supplies by preventing discharges of materials that can render a supply unusable for drinking even after reasonable levels of treatment. Table 1.1 shows the uses and sources of water in the United States from 1955 to 1975. It can be seen

TABLE 1.1   USES AND SOURCES OF WATER IN THE UNITED STATES[a]

USES AND SOURCES	1955	1960	1965	1970	1975
Off-channel water use (withdrawals)					
Public supplies	17	21	24	27	29
Rural domestic and livestock	3.6	3.6	4.0	4.5	4.9
Irrigation	110	110	120	130	140
Self-supplied industrial					
Thermoelectric power	72	100	130	170	190
Other self-supplied industrial	39	38	46	47	44
Total withdrawals	241.6	272.6	324.0	378.5	420
Water consumed by off-channel uses	—	61	77	87	96
Sources of water for off-channel uses					
Fresh ground water	47	50	60	68	
Saline ground water	0.65	0.38	0.47	1.0	
Fresh surface water	180	190	210	250	
Saline surface water	18	31	43	53	
Reclaimed sewage	0.2	0.1	0.7	0.5	
Water used for hydroelectric power	1500	2000	2300	2800	

[a] In billions of gallons per day. (Data sources: 1955–1970, Murray and Reeves (1972); 1975, Environmental Protection Agency (1981a).

that total use increased from approximately 240 billion to 420 billion gal day^{-1} in just 20 years. This increase in consumption is still continuing, and it is expected to increase at an even faster rate in the future. It should also be noted that most of the water currently used comes from surface water supplies. Surface water is the source most easily contaminated by human activity. It is the surface water systems, then, that must be carefully protected from contamination.

Water quality studies should determine which contaminant poses the most severe environmental problem. That is, a determination of the toxicity of many of the contaminants, or at least of model contaminants for specific groups, should be made. Heavy metals, organic pollutants, pathogenic organisms, etc., should all be understood with respect to their viability in natural systems and ability to cause environmental damage. With this information the threshold levels of the various contaminants can be determined and they can be regulated by imposing standards.

II. WATER QUALITY PROBLEMS

Although the purpose of this book is to supply information suitable for developing criteria for water quality, some previous information concerning water quality standards may be useful. Table 1.2 shows the frequency of violation of some of the Environmental Protection Agency's (EPA) water quality criteria. It can be seen that the criteria for many contaminants were exceeded at a very high frequency. For instance, iron was found in concentrations higher than the proposed standard 53% of the time. This indicates that either the water systems should be further purified or the standards should be evaluated with respect to their accuracy. These concepts will be discussed in Chapter 10, but it should be understood that large amounts of water quality data are necessary for setting realistic standards. It is obviously easier to alter the standard for a particular water contaminant than it is to purify the system. Thus, valid toxicological and chemical data must be available for accurate standards to be set.

Data have recently been collected on several water contaminants so that their environmental impact can be predicted. Many contaminants have now been monitored for a period of 10 years, so that long-term trends can be observed. However, the number of contaminants that have been monitored is relatively small, and few data are available for making overall environmental predictions. Figure 1.1 shows average concentrations of the pesticide DDT in U. S. surface water systems. Note that the levels of DDT decreased during the early 1970s, and they appear to be quite low at

TABLE 1.2   AVERAGE FREQUENCY OF OBSERVED VIOLATIONS OF THE
ENVIRONMENTAL PROTECTION AGENCY'S PROPOSED (1975) WATER
QUALITY CRITERIA AT NATIONAL WATER QUALITY SURVEILLANCE
SYSTEM STATIONS FOR 1974[a]

VARIABLE	AVERAGE FREQUENCY OF VIOLATION (PERCENTAGE OF ALL MEASUREMENTS)	NUMBER OF MONITORING STATIONS	NUMBER OF OBSERVATIONS
General variables			
Ammonia	11	52	844
Chloride	6	53	680
Dissolved oxygen	4	52	1120
Fecal coliform bacteria	67	47	907
pH	8	56	1168
Sulfate	18	53	645
Suspended solids	5	44	791
Trace metals			
Arsenic	<1	33	397
Cadmium	<1	36	454
Chromium	<1	39	463
Iron	53	49	744
Lead	16	35	471
Manganese	84	37	424
Zinc	44	46	577

[a] From Environmental Protection Agency (1981a).

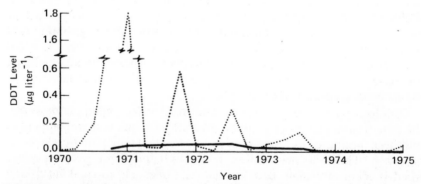

FIGURE 1.1   DDT LEVELS IN U. S. SURFACE WATERS: ···, raw data; ——,
2-year moving average.   [From Environmental Protection Agency (STORET).]

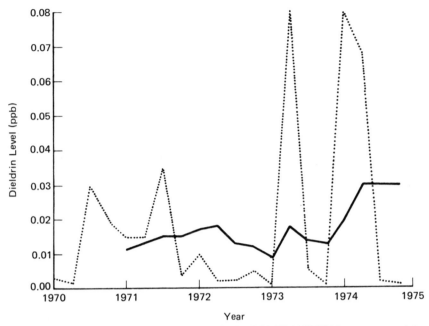

FIGURE 1.2   DIELDRIN LEVELS IN U. S. SURFACE WATERS: ⋯, raw data; ——, 2-year moving average. [From Environmental Protection Agency (STORET).]

present. This, of course, is due to a ban on the use of DDT in the United States beginning in 1972, which will be discussed in detail in Chapter 4.

Figure 1.2 shows the concentration of another pesticide, dieldrin, in U. S. surface waters for the same period. It can be seen that the average concentration of this pesticide in U. S. waters appears to be increasing, even though dieldrin has been essentially banned in the United States since the early 1970s. This is owing in part to the long residence time of the chemical in the environment, as well as the occurrence of an analogous pesticide, aldrin, which readily breaks down to form dieldrin.

Another water quality problem, which became apparent several years ago, is caused by the aqueous forms of mercury. Figure 1.3 shows mercury levels in U. S. surface waters, and there appears to be no discernible trend in the data. Even though mercury is known to be an environmental hazard, the curtailment of many of its uses has not appreciably lowered its concentration in U. S. surface waters.

Pesticides, polychlorinated biphenyls, and toxic metals such as mercury are known to cause environmental damage, yet only recent data have shown

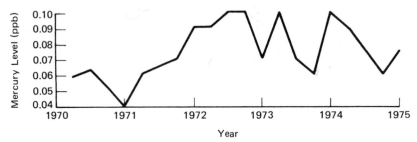

FIGURE 1.3   MERCURY LEVELS IN U. S. SURFACE WATERS.   [From Environmental Protection Agency (STORET).]

their ability to persist in water systems.   This is principally owing to the resistance of these elements to chemical degradation and hence their ability to be environmentally cycled.   Although there are undoubtedly thousands more compounds that pose the same or even greater potential for environmental harm, very little data are available on their behavior.   However, the families of pesticides, polychlorinated biphenyls, and toxic metals represent a fairly broad spectrum of compounds and can thus be used as models.   We shall use this approach throughout the text to show how new compounds and their potential for causing environmental damage can be estimated using information on closely related contaminants that have some kind of environmental data base.

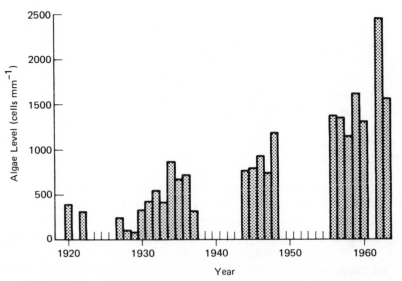

FIGURE 1.4   ALGAE LEVELS IN LAKE ERIE.   (From Davis, 1964.)

Some water quality problems are not characterized by direct toxicity to various components of the ecosystem. The environmental changes caused by these contaminants are more subtle, and they require a longer period of time for the effects to become visible. Such water quality problems occur when inorganic nutrients such as nitrogen and phosphorus are added to water systems. These nutrients act as fertilizers and increase productivity. The main problems caused by excess productivity are high levels of biomass and fluctuating oxygen concentrations, with the subsequent regeneration of many toxins. The overall enrichment process is called eutrophication, which will be discussed in Chapter 5.

Figure 1.4 shows some data on algae levels in Lake Erie. There has been a constant increase in the concentration of algae in the lake water over time, and many of the problems associated with the water quality of Lake Erie can be attributed to this increase in algal productivity.

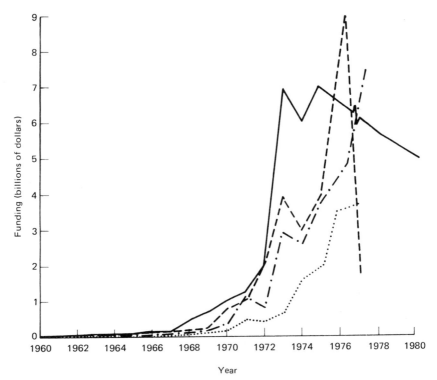

FIGURE 1.5 STATUS OF CONSTRUCTION GRANTS FUNDING FOR SEWAGE COLLECTION AND TREATMENT: ———, authorizations; –––, appropriations; — · —, obligations; ···, expenditures. (From Environmental Protection Agency, 1981b.)

As mentioned previously, the removal of various contaminants from waste is an expensive process and therefore decisions concerning treatment procedures must be guided by the contaminants' potential for harm in a natural system. Many constituents, such as dissolved organics, can be easily removed, thereby lessening the depletion of oxygen from the system. Certain other contaminants, however, are not easily removed by treatment processes and may pose serious environmental threats. Water quality relationships must be understood so that realistic waste treatment procedures can be developed.

The cost of wastewater treatment must also be considered, because in most instances it is extremely expensive. Figure 1.5 shows government expenditures for sewage collection and treatment in the United States. It can be seen that expenditures were relatively constant until the early 1970s, after which time they rapidly increased. These data do not include recent efforts to control non-point sources of pollution. The cost of containing stormwater runoff has always been prohibitive, but stormwater treatment is currently underway, and it is expected that enormous sums of money will be spent in this area in the future. Once again, it is imperative that we understand which pollutants pose the most serious threats to environmental health.

III. AN APPROACH TO DEVELOPING WATER QUALITY STANDARDS

In the following chapters we shall describe the fundamental microbiological and chemical processes that operate on various compounds in the aqueous phase. We shall see how organic and inorganic contaminants move through the environment, posing the threat of biomagnification. We shall also discuss the toxicities of well-studied pollutants and relate them to other materials that have similar toxicities. Microbiological processes involved in the movement of contaminants in natural systems will also be discussed, and in each case we shall attempt to show how microbiology and chemistry can be applied to make real engineering decisions that lead to improved water quality. In all cases particular compounds will be used as models, so that the potential for environmental damage of similar compounds can be predicted. There will be a strong emphasis on natural (including degradation rates) and physiochemical processes that must be understood to make engineering predictions. The culmination, then, will be an approach for setting realistic criteria and standards, because this is the only method for regulating the water quality of natural systems.

REFERENCES

Davis, C. C. (1964). Evidence for the eutrophication of Lake Erie from phytoplankton records. *Limnol. and Oceanogr.* **9,** 277.

Environmental Protection Agency (1981a). "Environmental Quality." U. S. Government Printing Office, Washington, D. C.

Environmental Protection Agency (1981b). "Review of the Municipal Waste Water Treatment Works Program." U. S. Government Printing Office, Washington, D. C.

Murray, R. N., and Reeves, E. B. (1972). "Estimated Use of Water in the United States in 1970," USGS Circular 676. U. S. Government Printing Office, Washington, D. C.

Chapter 2

Natural Environmental
Processes

I. INTRODUCTION

The overall theme of this book is water quality, which can easily be interpreted to mean water pollution. A strong effort has been made to limit use of the term "pollution" in this book because of its vagueness. In general, however, one can consider poor water quality to represent "polluted water." Because the effects of water quality can be noted on the overall environment, the term "ecology" is often used in conjunction with "pollution" when referring to water quality problems. In actuality, the science of pure ecology is different from the study of water quality, because there is little ecology in a highly perturbed area. However, ecological effects and principles can be helpful in describing the environmental effects of contaminated water.

Poor water quality can affect public health when toxic compounds build up in water systems. However, subtle changes are often more important in the long run. To understand how subtle changes develop due to human interference, we must first understand natural environmental processes.

Natural processes include fundamental ecological principles and energy transfer in ecosystems, which result in species stability. We can then superimpose perturbations caused by human activity in the form of decreasing water quality, onto these fundamental principles. In this way the environmental effects of the degradation of water quality may be better understood.

II. ENERGY TRANSFER IN ECOSYSTEMS

Perhaps the most fundamental concept in describing ecosystems is energy transfer. This concept has recently been recognized as the most basic model parameter in ecological systems. Even though it is a difficult parameter to quantify, the passage of energy through the different trophic levels of an ecosystem can be determined and the results related to water quality.

Other than certain radioactive elements, the only energy received by ecosystems comes from the sun. This energy is collected by a first trophic level, and passed through higher trophic levels in the ecosystem. Figure 2.1 is a schematic diagram of energy movement through a typical ecosystem. Primary producers collect solar energy, pass this energy onto the next level, usually a herbivore, and it is then passed onto the third level, usually a carnivore. Energy and matter are wasted at each level, and certain components of the ecosystem pass into a decomposer pool.

Radiant energy from the sun is absorbed by chlorophyll and other pigments in plants and converted into plant protoplasm; thus, photosynthesis is essentially a transformation of light energy into chemical energy. The radiation from the sun varies from short x rays to long radio waves, and of this radiant energy 99% has a wavelength in the range of 0.2 to 4 μm. This range includes the ultraviolet and the infrared, but 99% of the radiation lies in the visible region, that is, 0.38–0.77 μm. Therefore, it is not surprising that plant pigments such as chlorophyll absorb mainly at visible wavelengths.

The absorption of light by plants (photosynthesis) constitutes the primary producer level in an ecosystem. Equation (1) shows the overall stoichiometry of photosynthesis:

$$6CO_2 + 6H_2O \rightarrow C_6H_{12}O_6 + 6O_2. \tag{1}$$

Note that one molecule of sugar is formed at the expense of six molecules of carbon dioxide, producing six molecules of oxygen. Equation (1) suggests that photosynthesis generates only sugar, but obviously many other com-

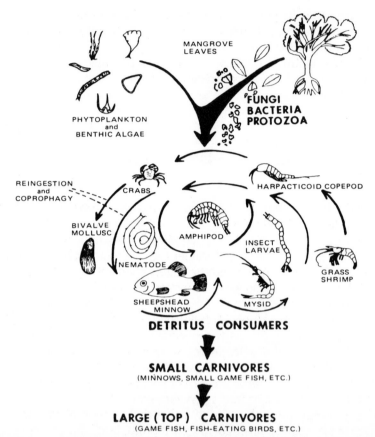

FIGURE 2.1   A DETRITUS FOOD CHAIN BASED ON MANGROVE LEAVES THAT FALL INTO SHALLOW ESTUARINE WATERS IN SOUTH FLORIDA.   Leaf fragments acted on by saprotrophs and colonized by algae are eaten and re-eaten (coprophagy) by a key group of small detritus consumers that, in turn, provide the main food source for game fish, herons, storks, and ibis.   (From "Fundamentals of Ecology," 3rd ed., by Eugene P. Odum. © 1971 by W. B. Saunders Company. Reprinted by permission of Holt, Rinehart, and Winston, CBS College Publishing.)

pounds are synthesized, including fats, acids, and oils, and each compound has a different percentage of carbon atoms.   Sugars are approximately 40% carbon by weight, whereas fats and oils are close to 80%.   This is of interest because the energy transferred between trophic levels is essentially dependent on the carbon content of the compounds involved.   It is also interesting to note that the carbon content of plant material is usually approximated as that of a carbohydrate, for which the percentage of carbon by weight is close to 40%.

The energy involved in photosynthesis is approximately 9.3 kcal g^{-1} of CO_2 fixed, assuming that the average compound formed is a carbohydrate. Plants, however, are not very efficient at converting light to chemical energy, because there is some heat loss during the energy transformation. In general, it can be shown that the energy conversion efficiency of plants varies between 1 and 2%, which leads to the typical energy pyramid for ecosystems.

The total productivity of ecosystems is also of interest. Table 2.1 shows primary production estimates for various habitats, and it can be seen that certain habitats appear to be very productive, whereas others do not. It is

TABLE 2.1   ESTIMATED GROSS PRIMARY PRODUCTION (ON AN ANNUAL BASIS) OF THE BIOSPHERE AND ITS DISTRIBUTION AMONG MAJOR ECOSYSTEMS[a]

ECOSYSTEM	AREA (10^6 KM^2)	GROSS PRIMARY PRODUCTIVITY (KCAL M^{-2} YR^{-1})	TOTAL GROSS PRODUCTION (10^{16} KCAL YR^{-1})
Marine			
Open ocean	326.0	1,000	32.6
Coastal zones	34.0	2,000	6.8
Upwelling zones	0.4	6,000	0.2
Estuaries and reefs	2.0	20,000	4.0
Subtotal	362.4	—	43.6
Terrestrial			
Deserts and tundras	40.0	200	0.8
Grasslands and pastures	42.0	2,500	10.5
Dry forests	9.4	2,500	2.4
Boreal coniferous forests	10.0	3,000	3.0
Cultivated lands with little or no energy subsidy	10.0	3,000	3.0
Moist temperate forests	4.9	8,000	3.9
Fuel-subsidized (mechanized) agriculture	4.0	12,000	4.8
Wet tropical and subtropical (broadleaved evergreen) forests	14.7	20,000	29.0
Subtotal	135.0	—	57.4
Total for biosphere (not including ice caps) in round figures	500.0	2,000	100.0

[a] From "Fundamentals of Ecology," 3rd ed., by Eugene P. Odum. © 1971 by W. B. Saunders Company. Reprinted by permission of Holt, Rinehart, and Winston, CBS College Publishing.

not surprising that tropical forests, densely cultivated areas, and certain estuary and swamp areas are very productive. It can also be seen that, in water systems such as the open ocean, productivity is very low, especially when only phytoplankton are considered. This table also gives insight into which systems are the most ecologically sensitive. Areas where productivity is high have a greater sensitivity to environmental perturbations. Thus, if we are considering the effects of water quality on various ecosystems, systems exhibiting high productivity should be of special interest.

Energy flow through ecosystems is an important parameter when considering the possible effects of poor water quality. Insight into the various trophic levels of an ecosystem can be gained by evaluating this energy transfer. After an initial loss of energy at the first trophic level, the efficiency of energy transfer up the food chain, or through different trophic levels of an ecosystem, varies from about 5 to 30%. The average gross ecological efficiency of transfer is only about 10% per trophic level. Thus, a given trophic level receives only 10% of the energy present in the lower trophic level, as shown by the following example.

Assuming 1500 cal m^{-2} day^{-1} of incident light strikes the plants of the first trophic level, only one-tenth of the energy would be incorporated into the plants, and only 10% of this would be synthesized as net plant productivity, so that we would end with 15 cal m^{-2} day^{-1} as the net primary productivity. Herbivores (the second trophic level) would then graze on this first trophic level and receive 1.5 cal m^{-2} day^{-1}. The third trophic level, carnivores, would receive approximately 0.15 cal m^{-2} day^{-1}.

Note that energy dissipates very quickly through trophic levels. This high energy loss dictates that most food chains in the environment be small, and in general we seldom find food chains containing more than three or four levels. For food chains containing more than three or four links, very little energy would be available for cellular maintenance and growth.

The largest energy transfer per unit takes place at the lower trophic levels, indicating that any perturbation at these levels would have a great influence on the rest of the food chain. This phenomenon is often referred to as biomagnification, which is the process by which a compound consumed at a lower trophic level in low concentration is passed on in greater concentration to the higher trophic levels.

Another consideration in analyzing energy transfer through an ecosystem is the rate of energy flow, and the classic method of analysis is to observe the transit time of energy in any trophic level. This is defined as the energy stored divided by the rate of energy flow. This transit time is quite variable in different systems, but it is usually high in aquatic systems. As an example, phytoplankton have energy transit times on the order of 9 to 18 days, whereas a rain forest that is equally if not more productive has transit times on the order of 25 years. This, of course, reflects the complexity of land plants compared with the simple structure of phytoplankton, but it also indicates that man must be very careful with systems that have high rates of energy transfer. Aqueous systems, although lower in productivity than terrestrial plant systems, are much more susceptible to small perturbations. This is because the rapid turnover rate leads to a correspondingly rapid build up of any toxic substance present in the system.

Using energy as a parameter for determining the extent of environmental perturbation in ecosystems is theoretically a reliable measure. It is a very difficult parameter to monitor, however, so quantification is not easily obtained. This is because water contamination can occur from many structurally different compounds, and the resultant environmental effects are difficult to evaluate. Therefore, the use of a general parameter that reflects the overall degree of perturbation in the system is desirable.

III. ECOLOGICAL SUCCESSION

An often-used parameter for monitoring the overall effects of water contamination is ecological succession. It has been observed in heavily contaminated water systems that the flora gradually change with time. The spectrum of change varies from very subtle species shifts to gross or major changes. In most cases the floral change caused by a given contaminant is consistently the same, and it is therefore possible to use this alteration as a parameter for determining water quality. Once again the ecological succession of a natural system must be understood so that a perturbed system can be compared.

Odum (1971) describes natural succession in an ecosystem as proceeding in the following manner:

(1) an orderly process of community change that is directional and predictable;

(2) a process resulting in part from modification of the environment by this community;

(3) a process culminating in a stable ecosystem.

More specifically, the development of an ecosystem in a natural environment would proceed in accordance with the following observations.

(1) The kinds of plants and animals change with succession, that is, there will be a flora and fauna associated with the early ecosystem or pioneer

system and a different flora and fauna associated with the final period of succession, called the climax.

(2)   The biomass and standing crop of organic matter will tend to increase with succession, that is, humus and dissolved organics will increase with time as a normal system evolves. This buildup of organic matter partly regulates further succession and reflects the biological activity that generally proceeds with ecological succession. It is this intimate relationship of dissolved organic matter and ecological succession that is very important to water quality.

(3)   The diversity of species in the community as a whole tends to increase with succession; thus, in a pioneer society or the early stages of a community, very few species are present. As the system matures and the ecosystem proceeds toward a climax community, the diversity tends to increase.

(4)   A decrease in net community production and a corresponding increase in community respiration occurs with succession. Therefore, a climax community can be characterized as an old, mature community that is stable because of its great species diversity, and there is a large organic structure with a balanced energy flow. This system, by definition, is very difficult to change, owing to its large diversity.

Armed with an understanding of natural evolution in ecological systems, the effects of poor water quality on aquatic ecosystems can be determined. When contaminants are introduced to a water system and the system changes in such a way that certain species can no longer exist, the system is forced backward with respect to natural evolution. Therefore, in many instances, such as when inorganic nutrients are introduced, water systems are driven from a stable to an unstable state.

Floral and faunal changes in water systems caused by sewage input, for instance, have long been used to monitor water quality. Figure 2.2 shows the classic scenario of the changes in flora and fauna in a stream receiving domestic sewage. When sewage is discharged into a stream, the more favorable species of fish (game, pan food, and forage fish) are precluded, and only a few species of objectionable forms (e.g., carp and catfish) proliferate. As the stream recovers, the natural flora and fauna may return to the point where the ecosystem is essentially the same as before the sewage input. Although these types of changes are qualitative at best, they can reflect the environmental changes due to a whole spectrum of contaminants. Thus, individual contaminants from a discharge may not have to be monitored. At least in a qualitative manner, species diversity can be used as an indication of environmental effects.

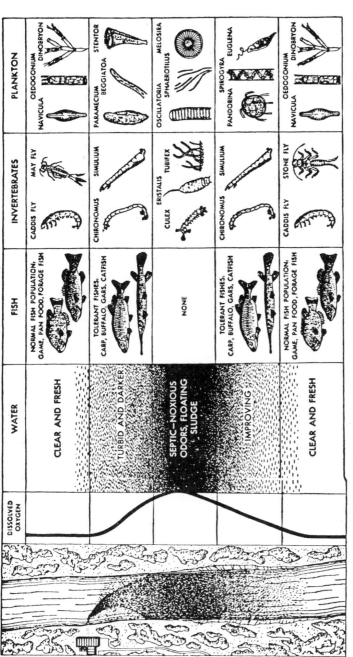

FIGURE 2.2 POLLUTION OF A STREAM WITH UNTREATED SEWAGE AND SUBSEQUENT RECOVERY AS RE-FLECTED BY CHANGES IN THE BIOTIC COMMUNITY. As the dissolved oxygen concentration in the water decreases, fishes disappear and only organisms able to obtain oxygen from the surface or organisms that are tolerant of low oxygen levels are found in the zone of maximum organic decomposition. When bacteria have reduced all of the discharged material the stream returns to normal. (From Eliassen, 1952. © 1952 by Scientific American, Inc. All rights reserved.)

IV.   THE DYNAMICS OF ORGANIC CARBON

One of the most important parameters that reflects ecosystem metabolism is the movement of organic carbon, and as shown in Section III, it is also used to define ecological succession. The production of organic matter in a natural ecosystem is indicative of development toward a mature state. The dynamics of the organic component are also important for determining the water quality of the system. The carbon cycle is discussed in detail in Chapter 5, so only the dissolved organic carbon fraction will be discussed here.

Organic carbon in aquatic systems is oxidized, or metabolized, predominantly by microorganisms. These microorganisms receive energy from the oxidation of dissolved organic compounds, and they are able to both synthesize protoplasm and use this energy to alter other compounds in the ecosystem. The utilization of organic carbon as a carbon source is called *heterotrophic activity.* The type of microorganisms most responsible for this activity are described in detail in Chapter 6.

The energy available from the oxidation of various organic compounds is variable, hence the energy received by microorganisms differs depending on the specific compound being metabolized. For example, methane (CH_4) yields only -24.4 kcal (electron mole)$^{-1}$, whereas formate ($HCOO^-$) yields -30.2 kcal (electron mole)$^{-1}$. This difference does not appear to be large, even though it is the maximum variation in energy transfer observed for organic compounds in nature.

It is often useful to model the microbial growth resulting from the uptake of dissolved organic carbon. As will be shown in Chapter 5, one of the most frequently used models is an enzyme kinetics model developed by Michaelis and Menten. When this model is modified to include a maintenance rate, the equation could be used for the production of heterotrophic bacteria at the expense of organic material. This relationship is given as

$$\mu = y_m k_m [S/(k_s + S)] - b, \tag{2}$$

where μ is the rate of bacterial growth, y_m the maximum yield of organisms per unit substrate consumed, k_m the rate of substrate consumption per unit mass of bacteria, S the concentration of rate-limiting substrate, k_s a constant that is equal to the substrate concentration that yields one-half the maximum growth rate, and b the coefficient of cell decay. In this equation k_m can be considered to be the electron transport rate, that is, the rate of oxidation of organics by oxygen. Recall that electrons are transferred during an oxidation–reduction process. McCarty (1972) has summarized the values for electron transfer rates given in the literature, and Table 2.2

TABLE 2.2 ELECTRON TRANSFER RATES

ELECTRON DONOR	ELECTRON ACCEPTOR	K_M (ELECTRON MOLE G^{-1} DAY^{-1})			E_A (KCAL/MOLE)
		6-15°C	16-25°C	26-35°C	
Glucose	O_2	0.12	0.35	1.0	18.2
Glucose	O_2	—	1.1	2.8	13.5
Maltose	O_2	—	0.9	1.9	11.7
Acetate	CO_2	—	0.6	1.1	9.9
Propionate	CO_2	—	1.0	1.0	—
Butyrate	CO_2	—	—	1.0	—
NH_4^+	O_2	—	—	2.9	—
NH_4^+	O_2	0.42	0.6	1.3	12.9
NH_4^+	O_2	0.7	1.1	2.8	16.8
NO_2^-	O_2	—	—	2.0	—
NO_2^-	O_2	0.53	0.7	1.2	9.8
NO_2^-	O_2	1.3	1.8	3.2	10.3
Fe^{2+}	O_2	—	—	0.6	—
Fe^{2+}	O_2	—	—	2.0	—
H_2	CO_2	—	—	3.1	—
H_2	SO_4^{2-}	—	—	2.4	—

shows a summary of transfer rates for different organic compounds. The values vary between 1 and 2 electron mole g^{-1} day^{-1} for the metabolism of both organic and inorganic compounds. Thus, even though the oxidation of various compounds in the environment appears to cover a broad spectrum, the actual electron transfer per gram of component per day is relatively constant.

Most of the organic carbon uptake in natural systems occurs through aerobic processes, that is, molecular oxygen is used by the microorganism to oxidize the organic compounds. There are also many processes that proceed anaerobically, without oxygen, but their overall contribution to the total organic pool is minimal. That is, only specific compounds are degraded principally by anaerobic processes in natural systems, because the rate of anaerobic oxidation of most organics is slow compared with aerobic oxidation.

Easily metabolized organics such as glucose are usually found in very low concentrations in natural systems. This is understandable in view of the ease with which they are utilized by microorganisms. The average concentration of glucose in most natural systems is approximately 10^{-7} M, and its turnover rate can be a good indicator of microbial activity. Vaccaro (1969) has observed turnover rates for glucose of about 42 h in the ocean, and

Siebert and Brown (1975) found turnover rates of less than 5 h in an estuary that was receiving contaminated effluent from a pulp mill. In the latter case they found that the uptake rate of glucose was approximately 0.24 h^{-1}.

It should be emphasized that many types of organic compounds are found in natural water systems. We have already mentioned glucose, but a whole range of compounds more difficult to break down than glucose are also present. The microflora responsible for the degradation of these compounds are in contact with a wide spectrum of compounds at any given time. It is therefore understandable that the microflora select certain of the more easily assimilated compounds to metabolize.

The concentration of total dissolved organics in natural systems varies from 1 to 20 mg l^{-1}. In general, bacteria metabolize the more easily assimilated compounds first and then attack the more refractory compounds. Figure 2.3 shows the type of growth curve expected when two different organic compounds are present in the same system. Note that growth proceeds at the expense of the easily metabolized glucose, then a lag in growth occurs, followed by subsequent growth utilizing the more recalcitrant galactose. This is called diauxic growth, and it shows the preference of a given microflora for a specific organic compound. Returning to Eq. (2), we see that equations that relate growth rate to organic carbon substrates must be defined as a single substrate and a single bacterium. That is, Eq. (2) could not model the removal of a group of organic compounds by a general population of microflora.

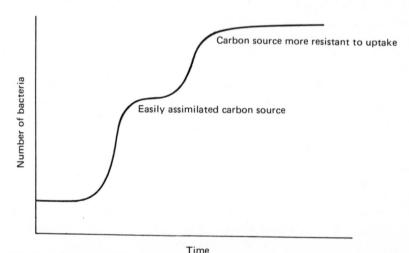

FIGURE 2.3   BACTERIAL GROWTH CURVE EXHIBITING DIAUXIC GROWTH FOR A SYSTEM CONTAINING A SUBSTRATE DIFFICULT TO UTILIZE AND A SUBSTRATE EASY TO UTILIZE.

We have seen that the build-up of dissolved organic matter in an ecosystem can be an indication of overall productivity. That is, as a system undergoes normal evolution, dissolved organic reserves increase and organisms evolve to use this nutrient pool as an energy source. The level of dissolved organic carbon has therefore also been used as an indicator of the overall health of an ecosystem. A system can be considered to be disturbed or to be moving away from natural evolution if organic matter builds up too rapidly. When this happens, it is usually due to the addition by humans of certain contaminants that accelerate overall productivity.

The dissolved oxygen system is integral to the dissolved carbon cycle. As mentioned previously, dissolved oxygen is used by microorganisms to stabilize organic matter in an ecosystem. Therefore, when the dissolved organic system is disturbed, so is the dissolved oxygen system. Because of the importance of dissolved oxygen for sustaining biological activity, we shall now address the dynamics of the dissolved oxygen system.

V. THE DISSOLVED OXYGEN SYSTEM

The aerobic stabilization of organic carbon by microorganisms consumes oxygen. A system that contains a large quantity of organic carbon therefore requires large amounts of oxygen for stabilization. In highly disturbed systems, this demand for oxygen may deplete all the available molecular oxygen, thereby precluding other biological processes that require oxygen. Because dissolved oxygen is required by aquatic organisms, its concentration has long been used as an indicator of water quality. Water quality standards have been set on dissolved oxygen levels, and the design of sewage treatment plants has long been based on preserving oxygen levels in receiving waters.

A. Biochemical Oxygen Demand

The utilization of oxygen by bacteria for the stabilization of organic matter can be approximated by a first-order differential equation. That is, the rate of oxygen used is a function of the amount of oxidizable material remaining in the system at any time:

$$dL/dt = -k_1 L, \tag{3}$$

where L is the amount of organic matter and k_1 the deoxygenation constant. In integrated form

$$L = L_0 \exp(-k_1 t), \tag{4}$$

where L_0 is the initial amount of organic matter (at $t = 0$) and L the amount of organic matter remaining after time t.   Figure 2.4 shows a plot of the first-order kinetics involved in the utilization of oxygen by bacteria.   Because the organic material in domestic sewage disrupts aqueous systems principally by consuming oxygen, it is desirable to relate the strength of domestic sewage to its ability to consume oxygen.   Sewage is considered to be oxidizable, and therefore it exerts a demand for oxygen.   This biochemical oxygen demand (BOD) is equivalent to L in Eq. (3) and has the units of milligrams per liter of oxygen.   The BOD of a sample of sewage can be considered a measure of the strength of the sewage.   The term L_0 in Eq. (4) actually reflects the initial strength of the waste, or its total potential to utilize O_2.

We generally do not know the BOD value for domestic sewage.   The classic method of determining this value is to measure the amount of oxygen consumed by a sample of sewage over a period of time.   This is defined by the term y, equal to $L_0 - L_t$, which is the BOD exerted at time t.   Again, the units of y are milligrams per liter of oxygen.   This equation can then be

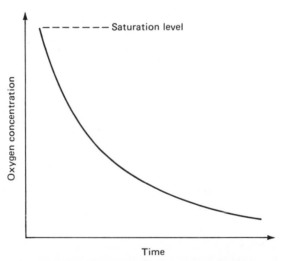

FIGURE 2.4   DISSOLVED OXYGEN CONCENTRATION VERSUS TIME FOR THE UTILIZATION OF OXYGEN BY BACTERIA.

substituted in Eq. (3) to yield

$$y = L_0(1 - e^{-kt}).\qquad(5)$$

This is the classic BOD equation, and Fig. 2.5 shows a plot of this function. Generally, a BOD analysis is run for 5 days, which is usually long enough to determine the first-stage BOD.

When using the BOD equation or attempting to predict the oxidation potential of domestic sewage, both L and k_1 must be determined. Therefore, solving this equation for the ultimate BOD as well as the rate constant requires statistical procedures.

The use of Eq. (5) to predict oxygen consumption by sewage is a very crude analysis. It does, however, reflect the true oxidizability of the waste by natural microflora. Figure 2.6 shows a more realistic picture of the microbial activity associated with the decay of domestic sewage. Bacteria grow very quickly on dissolved organic substrates, and after a short interval bacterial predators begin to feed on the heterotrophic bacteria. Both of these groups require oxygen, and the net organic carbon concentration of the system decreases rapidly because of this combined activity. The resultant oxygen demand curve (as shown in Fig. 2.6), however, represents a

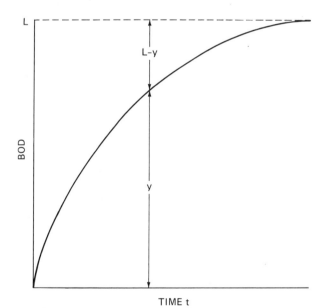

FIGURE 2.5   THE CLASSIC RELATION OF BIOCHEMICAL OXYGEN DEMAND (BOD) VERSUS TIME.

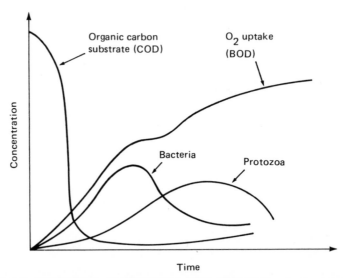

FIGURE 2.6   MICROBIAL ACTIVITY IN A BOD ANALYSIS.   As the organic substrate declines, oxygen uptake increases.   The primary microbial population is bacterial.   As the substrate declines, a secondary population of protozoa develops.   The protozoa utilize the bacteria as a substrate. (From Gaudy, 1972. © 1972 John Wiley & Sons, Inc.)

typical BOD curve.   Therefore, all of the different processes involved in the oxidation of organic matter by natural microflora result in a net level of oxygen uptake that is very close to that predicted by the BOD equation.

There are several other methods for measuring organic carbon besides the BOD test.   One is a wet chemical oxidation procedure in which a strong oxidant is added to the waste and the amount of oxygen consumed during the oxidation is calculated as the chemical oxygen demand (COD).   The sample can also be combusted and the organic carbon content measured as CO_2, or total organic carbon (TOC).   These analyses have both advantages and disadvantages compared with the BOD analysis.   The BOD analysis measures the fraction of organic material that can be oxidized by bacteria. This is important in natural systems for it is by this mechanism that organic carbon is actually degraded.   The COD and TOC analyses measure the amount of carbon present that can be oxidized by strong oxidants.   Generally COD and TOC values are higher than BOD values for the same waste sample.   This indicates that the indigenous microflora are unable to degrade certain parts of the waste, even though complete oxidation can be achieved by chemical or combustion methods.   The advantage is that the COD and TOC analyses can be run rapidly, whereas the BOD analysis takes 5 days to complete.

BOD analysis has been used for over 50 years to determine the oxygen requirement for the stabilization of domestic sewage, and most domestic sewage treatment plants have been designed to achieve BOD removal requirements. The use of BOD measurements for water quality predictions, however, requires that other parameters be considered. For instance, the net oxygen concentration in a surface water system is a combination of both oxygen loss (BOD reactions) and oxygen input from various physical and biological processes.

B. Oxygen Balance in Natural Systems

When organic matter is discharged into a water system, oxygen is consumed. The prediction of oxygen consumption and net oxygen concentrations in aqueous systems is a function of both aeration and oxygen loss from BOD reactions. The mathematics of this process were developed years ago; the first model being proposed by Streeter and Phelps (1925). They first proposed a model that would predict oxygen concentrations in the Ohio River as affected by domestic sewage inputs. The original model was called the "oxygen sag curve" and included only oxygen removed by BOD reactions and oxygen added by physical aeration. Since that time other terms have been added to the model. We shall review the most applicable form of the model in current use today. This model attempts to predict the dissolved oxygen concentration or deficit at any distance downstream from a point source of domestic sewage.

Oxygen is dissolved in surface waters by physical aeration, and the rate of input is a function of the oxygen deficit in the water system, that is, the difference between the actual concentration of oxygen and the theoretical concentration at saturation. This relationship is

$$R_1 = k_2(C_s - C), \tag{6}$$

where R_1 is the rate of aeration in milligrams per liter per hour, k_2 the aeration coefficient in h^{-1}, C the concentration of dissolved oxygen in solution in mg liter^{-1}, and C_s the saturation concentration of oxygen in solution in mg liter^{-1}. The rate at which oxygen is removed from the water system is a function of the BOD, that is, the rate is dependent on the amount of oxidizable organic matter present:

$$R_2 = -k_1 L, \tag{7}$$

where L is the BOD remaining after time t in milligrams per liter and k_1 the deoxygenation constant in h^{-1}. In addition to oxygen input from physical aeration, oxygen can also be added to a system by photosynthesis. However, plant material respires, thereby removing oxygen from a system as well as adding it. Therefore, another rate term should be added to reflect flow of oxygen to or from a system caused by plant activity. We shall write this term as

$$R_3 = -S_r, \tag{8}$$

where S_r is the rate at which the concentration of dissolved oxygen changes as a result of photosynthesis and respiration in milligrams per liter per hour. The negative sign indicates that the rate of oxygen removal is greater than the rate of oxygen addition. If a system is operating such that more oxygen is added by photosynthesis than removed by respiration, the sign would be positive.

A model can now be constructed that reflects the oxygen balance of the stream as influenced by a point source of a dissolved organic material such as domestic sewage. The model will be constructed using the mass balance relationships described in detail in Chapter 9. The total change over time in oxygen concentration in a stream reach is

$$dC/dt = k_2(C_s - C) - k_1L - S_r. \tag{9}$$

It is convenient to express this total derivative in terms of its partial space and time derivatives:

$$\frac{\partial C}{\partial x} V_x + \frac{\partial C}{\partial t} = k_2(C_s - C) - k_1L - S_r, \tag{10}$$

where we are only concerned with the x component of the velocity.

We further assume that the volumetric flow rate and cross-sectional area of our reach remain constant, so that $V_x = Q/A$, where Q is the volumetric flow rate in cubic meters per second and A the cross-sectional area of the stream in square meters. Finally, we assume that the reach is completely mixed (that diffusion is small); therefore,

$$\frac{\partial C}{\partial t} = \frac{-Q}{A} \frac{\partial C}{\partial x} + k_2(C_s - C) - k_1L - S_r. \tag{11}$$

Owing to the dependence of oxygen solubility on salinity and temperature, Eq. (11) is often written in terms of oxygen deficit. The oxygen deficit D is defined as the difference between the oxygen concentration at saturation and the concentration at any time t:

$$D = C_s - C. \tag{12}$$

After substitution, Eq. (11) becomes

$$\frac{\partial D}{\partial t} = \frac{-Q}{A} \frac{\partial D}{\partial x} - k_2D + k_1L + S_r. \tag{13}$$

At steady-state conditions, $\partial C/\partial t = 0$ and $\partial D/\partial t = 0$; Eqs. (11) and (13) then reduce to

$$\frac{Q}{A}\frac{\partial C}{\partial x} = k_2(C_s - C) - k_1 L - S_r \tag{14}$$

and

$$\frac{Q}{A}\frac{\partial D}{\partial x} = -k_2 D + k_1 L + S_r. \tag{15}$$

These equations can be integrated to yield a spatial representation of oxygen concentration in a stream reach (see Chapter 9).

VI.   CONCLUSION

In summary, we have seen how the formation and destruction of the dissolved organic content of water systems are related to ecological concepts. These processes are integral to basic water quality relationships, because dissolved oxygen has been used for many years as an indicator of change in water quality. Recent advances in technology, however, have led to the discharge of more sophisticated toxins into our water systems, hence a general dissolved oxygen model cannot adequately reflect the subtle changes now taking place in aquatic ecosystems. Many refractory organics as well as inorganic materials can undergo biomagnification without resulting in changes in an ecosystem that can be discerned by measuring changes in the concentration of dissolved oxygen. Therefore, we must now investigate the specific interactions that take place with individual pollutants. The following chapters will define the environmental movement of these contaminants and describe the subtle ecological changes that occur because of their presence.

REFERENCES

Eliassen, Rolf (1952). Stream pollution. *Sci. Am.* **186**(3), 17–21.
Gaudy, A. (1972). Biochemical oxygen demand. *In* "Water Pollution Microbiology" (R. Mitchell, ed.), pp. 305–332. Wiley, New York.

McCarty, P. L. (1972).   Energetics of organic matter degradation.   *In* "Water Pollution Microbiology" (R. Mitchell, ed.), pp. 91–118.   Wiley, New York.

Odum, E. P. (1971).   "Fundamentals of Ecology."   W. B. Saunders Co., Philadelphia, Pennsylvania.

Siebert, J., and Brown, T. J. (1975).   Characteristics and potential significance of heterotrophic activity in a polluted fjord estuary.   *J. Exp. Mar. Biol. Ecol.* **19,** 97–104.

Streeter, H. W., and Phelps, E. B. (1925).   "A Study of the Pollution and Natural Purification of the Ohio River.   III.   Factors Concerned in the Phenomena of Oxidation and Reaeration," Public Health Bulletin No. 146.   U. S. Public Health Service.

Vaccaro, R. F. (1969).   The response of natural microbial populations in sea water to organic enrichment.   *Limnol. and Oceanogr.* **14,** 726–735.

Chapter 3

Toxic Metals and Water Quality

I. INTRODUCTION : M O

Toxic metals are relatively important items on the list of water quality problems. Certain elements resist treatment and therefore can be long lived in the environment, allowing time for biological processes to incorporate metals into biomass and for biomagnification to occur. Although the concentration of a particular metal in a waste discharge may be small and the discharge of short duration, the concentration can be magnified many times by the natural biota. When this biomass is consumed at higher trophic levels, by either humans or other animals, the consumer can receive a large dose of the absorbed metal. The environmental residence time of metals is the most important concept in this chapter.

There is a difference between metals and recalcitrant organics with respect to their ability to persist in the environment. Organic compounds such as pesticides, which will be discussed in Chapter 4, have a unique and ordered chemical structure. The atoms in these structures are arranged in well-defined systems of chemical bonds that can be broken with the small amounts of energy within the capacity of natural systems. Organic compounds such as DDT, therefore, will ultimately be broken down, even though they can persist for long periods of time. The character and therefore the effect of an organic compound is dependent on its structure; once that structure is destroyed, the effect changes. Yet, there is a difference in the environmental residence time of organics compared with metals, which require energy inputs at the molecular level to be transformed. This seldom occurs in nature. Metals persist indefinitely in one form or another; therefore, it can be considered that metals pose a much greater threat to the environment than do even the "persistent" organics.

Metal toxicity in an industrial setting has been a concern for many years. As early as 1941 it was recognized that mercury caused hattershakes, a disease common to workers in factories where felt hats were made. This was owing to the use of mercuric nitrate in processing the felt. Since then other metals, e.g., cadmium, zinc, and lead, have been implicated as causative agents in nervous disorders, bone weakening, and heart disease.

Lead poisoning has also been recognized as a public health problem that can occur in many ways. Infants who ingested lead-based paints, for example, became inflicted with a type of lead poisoning that is very difficult to diagnose in its beginning stages owing to the slow development of definitive symptoms. This was especially true in children, and many of the symptoms of lead poisoning were confused with mental retardation or personality disturbances. In the same manner, over 100 cases of mercury poisoning in Minnamata Bay, Japan were not correctly diagnosed from 1953 to 1960.

The metals that are commonly involved in biological activity are the lighter elements in the periodic table, that is, elements with atomic numbers less than about 40. These are also the metals that form more soluble compounds and hence are widely distributed in nature. The heavier metals are not generally involved with the activities of living systems and can therefore be considered to be water pollutants.

II. PROPERTIES AND BEHAVIOR OF TOXIC METALS

This section will describe the behavior of those metals that have been studied in detail, and an attempt will be made to define sources, environmental damage, and occurrence in natural systems. There are many metals that have extreme potential for environmental degradation that will not be discussed, but we assume that these metals follow models similar to those described.

A. Mercury

Mercury is one of the least abundant elements in the earth's crust, ranking seventy-fourth out of a total of 90. The principal mercury-containing ore mined in the world is cinnabar. Over the ages mercury has been used for many purposes, including extensive use in the medical profession. Physicians and surgeons have used mercury compounds as bactericidal agents for hundreds of years. The use of mercury, even though an extremely common practice, was never evaluated closely for health effects. With respect to environmental effects, very little attention was paid to mercury compounds in water systems in the United States until 1970, even though Minnamata disease in Japan had already been attributed to discharges of mercury.

In 1967 the Swedish Medical Board found it necessary to prohibit the sale of fish from certain Swedish rivers because of high levels of mercury. It was not until 1969, when researchers found high levels of mercury in Canadian fish, that international attention was focused on the occurrence of mercury in natural systems. Subsequent studies found that there were significant levels of mercury in some of the higher pelagic fish such as tuna and swordfish.

1. Toxicological Effects

Metallic mercury is of little interest from the point of view of health, because relatively large quantities can be consumed without severe health effects. However, mercury vapor that sublimes from elemental mercury at room temperature can cause acute toxicological effects in the lungs, often leading to the necrosis of pulmonary tissue.

The most important forms of mercury from a water quality perspective are the salts. The best-known soluble salt of mercury is mercuric chloride $(HgCl_2)$. This compound is extremely poisonous to humans and has been used in many suicides. Mercurous chloride $(HgCl)$ is less dangerous in water than mercuric chloride, mostly because of its lower solubility. We shall see in the next section how the solubility of these salts generally governs their toxic effects on humans.

It is the organic forms of mercury, however, that have commanded the greatest attention with respect to water quality. Alkylmercury salts, which have caused a great number of human poisonings, include the well-known methylmercury forms. Almost all alkylmercury compounds behave in the same manner when absorbed by the human body. The victim receives permanent brain injury, resulting in weakness and often progressing to paralysis with subsequent loss of vision and disturbed cerebral functions. A number of cases have been documented in Japan in which healthy females have given birth to offspring with brain damage. This is due to the ability of certain forms of mercury to cross placental membranes and affect the fetal brain.

The ability of mercury to be transformed through the biological food chain is essentially a function of how long the element is retained by any particular level. Goldwater and Clarkson (1972) published a list of the residence times of mercury in various animals (see Table 3.1). It can be seen that the half-life of mercury in animals varies over a wide range. For instance, a mouse tends to discharge half of its total mercury intake in a little over a week, whereas certain invertebrates and fish require over 3 years to release half of a single mercury dose. It is the long retention time of an element in biological entities that can lead to biomagnification and subsequent health effects.

2. Chemical Ecology

The ecology of mercury compounds has been outlined by Fagerstrom and Jernelov (1972), who showed that mercury can occur in the environment in several forms, as is the case with many other heavy metals. Mercuric sulfide (HgS) has the lowest solubility of any mercury compound ($K_s = 10^{-53}$), hence it does not interact to any great extent in environmental systems. Elemental mercury can be oxidized to bivalent mercury by natural processes. In like manner mercury in its $+2$ valence form can also be reduced to elemental mercury if oxidation–reduction conditions are suitable. This process appears to occur in many sediment systems and can be catalyzed by several groups of bacteria.

TABLE 3.1 BIOLOGICAL HALF-LIFE OF METHYLMERCURY IN DIFFERENT ANIMALS[a]

ANIMAL	HALF-LIFE (DAYS)
Mouse	8
Squirrel monkey	65
Seal	500
Rat	16
Poultry	25
Molluscs	481–1000
Crab	400
Pike	640–780
Flounder	700–1200
Eel	910–1030

[a] From Goldwater and Clarkson (1972).

Mercury in its bivalent oxidation state readily combines with either inorganic or organic material. It has a great affinity for certain organic substances and will readily form charged organic material. In addition, bivalent mercury forms two distinct types of inorganic compounds. The first has been referred to as the silica type, which involves a mercury–silica complex, but this compound does not appear to be oxidized or reduced in natural systems. The other bivalent mercury–inorganic complex has been designated as the theromanganese type, which can be oxidized or reduced within the redox potential range of natural systems.

The solubility of various inorganic species of bivalent mercury is important to consider at this point. Hahne and Kroontje (1973) calculated solubilities for various mercury hydroxides, and Fig. 3.1 shows a plot of the fraction of various mercury hydroxides as a function of pH. Note that the relatively insoluble mercuric hydroxide is formed at pH values greater than 6, which are normally encountered in natural systems.

Figure 3.2 shows the effect of chloride concentration on mercury. Chlorides are present in most water systems, and they often regulate the complexation of various species of mercury. There is generally a competition in the inorganic phase between chloride and hydroxyl complexation. Mercury is unique in that it complexes with chloride ion at much lower concentrations than other toxic metals. An example of how the inorganic species of mercury interact is presented in Fig. 3.3, which shows calculated

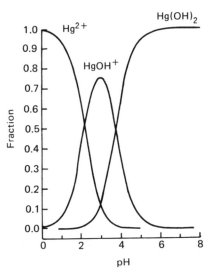

FIGURE 3.1 RELATIVE DISTRIBUTION OF SPECIES OF MERCURY COMPOUNDS VS. pH. (From Hahne and Kroontje, 1973.)

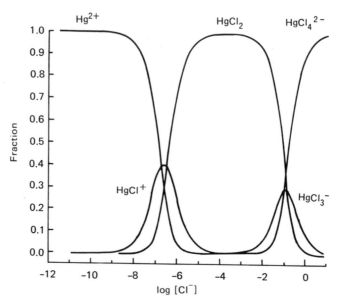

FIGURE 3.2   DISTRIBUTION OF SPECIES OF MERCURY COMPOUNDS VS. CHLORIDE ION CONCENTRATION.   (From Hahne and Kroontje, 1973.)

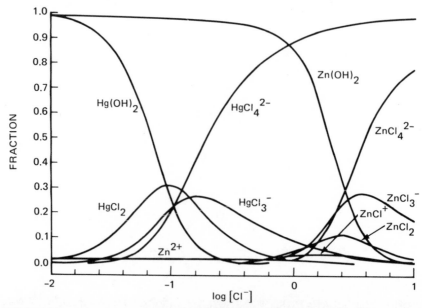

FIGURE 3.3   RELATIVE DISTRIBUTION OF HYDROXIDE AND CHLORIDE COMPLEXES OF MERCURY AT pH = 8.5.   (From Hahne and Kroontje, 1973.)

distributions of mercury and zinc at a pH of 8.5. As the chloride concentration is increased, the hydroxide forms of mercury become less common, and the predominant species becomes either $HgCl_3^-$ or $HgCl_4^{2-}$.

We have seen so far that the solubility of the inorganic forms of mercury relate directly to the ability of mercury to inflict environmental damage, that is, the more soluble the compound in aqueous systems, the more likely it is to enter biological food chains and become a water quality hazard. We have also seen that there is an interaction between hydroxide and chloride ions that regulates the solubility of mercury in natural systems.

3. Formation of Methylmercury

Probably the most studied and most important form of mercury from a water quality perspective is methylmercury $((CH_3)_xHg)$. Mercury can be methylated by microorganisms during biological processes. Methylation of mercury appears to occur in both anerobic and aerobic systems at rapid rates. Either methylmercury (CH_3Hg) or dimethylmercury $((CH_3)_2Hg)$ is formed during this process, and the relative distribution of the different species is dependent on pH. Figure 3.4 shows the relative amounts of dimethylmercury and methylmercury formed in sediments as a function of pH. At pH values greater than 8 or 9 almost all of the methylated mercury is in the form of dimethylmercury, and at pH values less than 8 methylmercury is the favored species. This distribution is important, because dimethylmercury is quite volatile and would probably evaporate into the atmosphere at the higher pH values. However, methylmercury remains

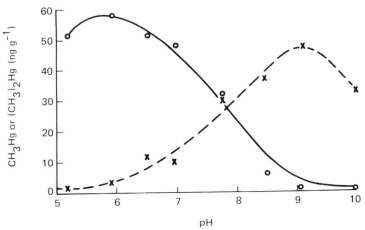

FIGURE 3.4   FORMATION OF METHYL (o) AND DIMETHYL (✕) MERCURY IN SEDIMENTS VS. pH. (From Fagerström and Jernelöv, 1972. © 1972 Pergamon Press, Ltd.)

incorporated in the water system and accumulates in living organisms. The potential for accumulation of methylmercury appears to be predominant at pH levels lower than 7.5.

Several studies have shown that methylmercury can also be degraded by bacteria. Spangler *et al.* (1974) have shown that out of a total of 207 bacterial isolates, 30 were found capable of demethylating methylmercury and could tolerate concentrations of mercury up to $0.5 \, \mu g \, ml^{-1}$. The processes studied were aerobic, producing methane as a degradation product. It can be seen that mercury can be methylated or demethylated by bacterial activity, and different forms of methylmercury can occur in different environmental situations.

It is usually the methylated form of mercury that is found in the tissue of higher carnivores, owing to the fact that this form is readily taken up by organisms. We have also seen that mercury in any form is readily methylated by common groups of bacteria, making it available for biological uptake. Therefore, a direct discharge of methylmercury to an aqueous system is not required for a buildup of the methylated forms. Methylmercury can also be demethylated, and it is probably this process that keeps the level of methylmercury in sediments at the low levels observed in most natural systems.

B. Lead

As does mercury, lead occurs in natural deposits throughout the world. Lead is usually found as a sulfide, oxide, or carbonate ore. Like mercury, lead has for centuries been used for various industrial purposes: as a component of some paint pigments, in the manufacture of various chemicals, and as a gasoline additive. The large amounts of lead previously used in gasoline were probably the prime source of environmental contamination.

The $Pb-H_2O$ system usually determines which form of lead will be found in normal environmental situations. Redox equilibria determines that Pb^{2+} cannot exist in appreciable concentrations in a natural system at equilibrium. The normal form would be the solid, PbO, or, at higher redox potentials, PbO_2.

In systems containing carbonate ion, which is common in natural systems, lead carbonate compounds are formed. As with other metals, there is an interaction between the forms depending on the pH and redox potential of the system. Stumm and Morgan (1970) constructed a typical $p\epsilon-pH$ diagram for the $Pb-CO_2-H_2O$ system (see Fig. 3.5). This diagram shows

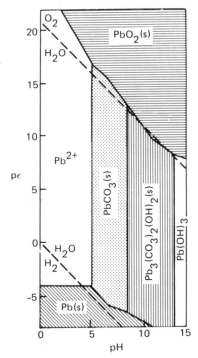

FIGURE 3.5 REDOX EQUILIBRIA FOR A Pb–CO_2–H_2O SYSTEM. (From Stumm and Morgan, 1970. © 1970 John Wiley & Sons, Inc.)

that the lead carbonate phase is thermodynamically the most stable in the neutral pH and middle pϵ ranges. At high pϵ situations, that is, highly oxidized systems, lead oxide compounds are favored, whereas for low pϵ values, elemental lead is the favored species.

It should also be noted that lead readily forms additional solid phases with other anions such as chloride, sulfate, and phosphate. Consequently, many other forms of lead can occur in natural systems. The schematic of the Pb–CO_2–H_2O system in Fig. 3.5, however, is a representative model of how certain lead species vary as a function of pH and pϵ.

Lead is generally an insoluble metal, hence in natural water systems it occurs in extremely low concentrations. When lead in aquatic systems first became a water quality concern, a study was undertaken to determine lead concentrations in U. S. drinking water supplies. Data from that study showed that approximately 14% of the samples tested from 1963 to 1965 contained more than 19 μg l^{-1} of lead (Symposium on Environmental Lead contamination, 1966). Rainwater in industrial or roadside areas, on the other hand, can contain concentrations as high as 50 μg l^{-1} of lead.

Lead can be introduced to a water system by several mechanisms in mining areas where limestone and galena ores are found. Natural water systems can contain lead concentrations as high as 0.8 μg l^{-1}, usually introduced by leaching from lead chromate paints or old pipe installations or from acid dissolution of solders. Lead can also be removed from soils by absorption. Bittell and Miller (1974) showed that lead has a very high specific exchange rate for the three clays montmorillonite, illite, and kaolinite. They also showed that of three metals tested — lead, cadmium, and calcium — lead was absorbed preferentially by a factor of two or three over the other metals.

Perhaps the most important input of lead to environmental systems in the recent past has come from automobile emissions. It was mentioned earlier that large amounts of lead were once used as antiknock compounds in gasoline. Although it is true that for the last few years the use of lead as a gasoline additive has diminished, there are still substantial inputs from this source. It appears that lead from gasoline combustion is emitted to the atmosphere principally in the form of particulate matter. These small particles fall to the earth and become available for incorporation into biological systems. It has been shown that organisms close to highways can contain high levels of lead. It has also been shown that individuals inhaling small lead particles can receive large doses of lead. This is owing to the fact that one of the most efficient methods of contracting metal poisoning in humans is by the inhalation of metal particles.

We have seen that lead can be introduced into aquatic systems by several human activities. Lead remains relatively insoluble in all its forms and does not occur in large concentrations in aquatic systems unless some sort of contamination is present. It should also be noted that lead does not undergo methylation, as does mercury. As a potential water contaminant, then, lead does not appear to pose as severe a threat as does mercury or any metal that is capable of being methylated. It has been shown, however, that lead is dispersed to the environment principally by the combustion of lead compounds in gasoline. These emissions tend to be in the form of small particulate matter that can be easily assimilated by humans. For this reason, lead appears to generate health concerns different from those associated with mercury.

1. Health Aspects

Lead poisoning causes several well-known but nonspecific illnesses in humans. Recent medical evidence suggests that the various syndromes exhibited are a function of time and dose relationships associated with lead intake. Plumbism, the earliest manifestation of lead poisoning, was originally taken for anemia. In the human body, lead prevents the uptake of iron, hence preventing the formation of hemoglobin. Victims lose red blood cells and exhibit all of the syndromes of anemia, that is, pale skin, fatigue, irritability, and mild headaches.

Perhaps the most severe type of lead poisoning is referred to as acute encephalopathy. In this case the victim usually undergoes convulsions, lapses into a coma, and in severe cases dies within 48 h. This situation occurs only for very high levels of lead intake. Many children have suffered from this disease, and often in situations where it was not fatal, a large percentage of the victims sustained permanent brain damage.

2. Biomagnification

As for most metals the ability of lead to cause environmental harm rests on its tendency to be absorbed at lower trophic levels and to subsequently undergo biomagnification. We have seen that this indeed occurs with mercury, especially when mercury becomes methylated. However, as previously mentioned, lead does not appear to be methylated in natural systems and hence does not move as readily up the food chain. A study by Skaar et al. (1973) showed that mosses growing beside heavily traveled highways accumulated lead and in some instances contained levels up to 50 ppm dry weight. It has been mentioned that particulate lead from automobiles is deposited very close to highways and can be readily assimilated by the flora. The movement of lead up the food chain from vegetation close to highways has not yet been determined. Therefore, it is not clear whether this is an important mechanism in environmental situations.

Several studies of aqueous systems have been performed to evaluate the increase in lead concentration of various aquatic animals caused by lead discharges. Valiela et al. (1974) have determined the lead content of certain bivalve molluscs as a function of sewage sludge enrichment that contained relatively high levels of lead. They showed that none of the bivalves exhibited an increase in the concentration of lead in body tissues as a result of the fertilization. They also showed that virtually all of the lead entering the marsh system from the sewage was retained in the sediments. It is conceivable, however, that plant material that had accumulated lead could be washed from the system and hence transfer the lead to other locations.

Even though lead is considered to be an environmental contaminant and large amounts are being introduced by human activities, it appears that very little biomagnification of lead occurs. Studies to date have shown a minimal increase in the level of lead through trophic chains, which is due in part to the limited environmental movement of lead.

C. Zinc and Cadmium

1. *Environmental Distribution*

Zinc and cadmium shall be treated together in this section because their chemistry is quite similar. In fact, three elements — mercury, cadmium, and zinc — are all similar in chemical structure. Environmentally there is quite a difference, however, because mercury is extremely toxic, whereas zinc is required in large concentrations for normal biological metabolism. Cadmium is located between zinc and mercury relative to toxicity. Zinc and cadmium occur together in nature and are difficult to separate. It is only their different boiling points that allows for the industrial separation of the two elements.

Zinc is required in large concentrations by most organisms, and humans, for instance, require approximately 6–15 mg day^{-1} of zinc. The only metal required in a larger concentration by humans is iron. Indeed, it seems that the body is more sensitive to a lack of zinc than to excess amounts of zinc. When diets become deficient in zinc, dwarfism can occur. On the other hand, excess intake of zinc has been shown to be beneficial for such things as healing wounds. It therefore appears that zinc should not normally pose a severe water quality problem.

Cadmium retains many of the problems of its closely allied metal, mercury. It readily forms complexes in natural systems with ammonia and cyanide, and it forms organic metal complexes, which is the major water quality problem associated with mercury compounds. Although cadmium can also be methylated, the carbon – cadmium bonds are relatively unstable compared with those of mercury. Therefore, organic cadmium compounds do not remain for long periods of time in the environment.

TABLE 3.2 RESIDENCE TIMES AND
CONCENTRATIONS OF SELECTED
ELEMENTS IN SEAWATER[a]

ELEMENT	ppm	RESIDENCE TIME (YEARS)
Cd^{2+}	0.00011	500,000
Cu^{2+}	0.003	50,000
$HgCl$	0.00003	42,000
Pb^{2+}	0.00003	2,000
Zn^{2+}	0.01	180,000

[a] From Bowen (1966).

TABLE 3.3   OCCURRENCE OF SELECTED
ELEMENTS IN U. S. FRESH WATERS[a]

ELEMENT	FREQUENCY OF OCCURRENCE (%)		CONCENTRATION (μg/l)	
	RAW	FINISHED	RAW	FINISHED
Cadmium	2.5	0.2	9.5	12.0
Copper	75.0	65.0	15.0	43.0
Lead	19.0	18.0	23.0	34.0
Zinc	77.0	77.0	64.0	79.0

[a] From Kopp (1969).

Cadmium is not commonly involved in environmental processes, but it is used in relatively large amounts in industry. Cadmium behaves much like lead in aqueous systems, with one major difference — the mean residence time. Lead readily precipitates because of its low solubility and becomes bound in soil matrices. Cadmium does not appear to behave in this manner. Table 3.2 shows calculated residence times for various elements in seawater. Note that the longest-lived metal appears to be cadmium, with a mean residence time of abut 500,000 yr. This suggests that cadmium has the potential of building up in the environment and causing environmental damage. On the other hand, lead has a short residence time and zinc is second to cadmium with respect to longevity in seawater.

Kopp (1969) examined the cadmium content of 15,000 raw and 380 finished waters in the United States. Table 3.3 shows the frequency of certain concentrations of cadmium in the water systems. Note that cadmium is usually found in low concentrations and rarely found in concentrations greater than the EPA drinking water standard of 0.01 mg l^{-1} in natural systems.

2.   Health Aspects of Cadmium

It has been known for some time that cadmium is an extremely toxic element and probably more lethal than most other metals when breathed as a fume or gas in industrial situations. Unfortunately, information concerning the metabolic effects of cadmium on humans is sparse. It does appear, however, that cadmium can replace zinc in several physiological processes in the body, resulting in bodily disorders. The human body is apparently unable to differentiate between zinc and cadmium; hence, they are both assimilated when both are present. However, zinc is an essential element for human functions, whereas cadmium appears to be toxic.

Medical studies have shown that cadmium poisoning results in severe vomiting for doses of cadmium in excess of 15 mg. This is a rather large dose; therefore, cadmium should not be a severe health hazard, owing to its low concentrations in aqueous systems. There has been considerable interest in a Japanese disease called itai-itai (ouch-ouch), which appears to effect women between 50 and 60 years of age and is characterized by pains in the joints and bones. This disease was originally attributed to cadmium poisoning resulting from mining activities. It has recently been shown, however, that other factors such as basic nutritional deficiencies in vitamin D and calcium were probably the causes of the disease.

3. Biomagnification

As with other metals, the potential for cadmium to undergo biomagnification and become a health hazard depends in part on its ability to be taken up by plants. Table 3.4 shows the concentrations of some elemental metals in soil and plants. It is shown that zinc and lead are not readily taken up by plants but are mostly retained in the soil. However, cadmium concentrations are at least tenfold higher in plant tissue than in soil. The movement of cadmium from this lower trophic level up the food chain, however, has not been proven. This finding agrees with Bittell and Miller's (1974) work, which showed that cadmium is exchanged in clays at about the same rate as calcium. We recall here that lead is adsorbed onto clays to a much greater extent than either cadmium or calcium. Therefore, cadmium has a lesser affinity for ion exchange in soil systems than does lead. Once again referring to the work of Valiela et al. (1974) on the fertilization of salt

TABLE 3.4   SELECTED ELEMENTAL
METAL CONCENTRATIONS IN SOIL AND
PLANTS[a]

| | CONCENTRATION (ppm) | | |
ELEMENT	SOIL	PLANTS	PLANT–SOIL RATIO
Cadmium	0.06	0.64	10
Zinc	50	32	0.6
Lead	10	4.5	0.45
Copper	20	9.3	0.45

[a] From Cannon (1969).

marshes with sewage sludge, it was shown that the cadmium concentration in all the shellfish tested increased significantly after fertilization. The concentration of cadmium in marsh sediments also increased because of sewage enrichment. It appears that cadmium readily undergoes biomagnification, and it is this phenomenon that identifies cadmium as a water quality problem comparable to mercury.

III. CHEMICAL AND BIOLOGICAL ALTERATIONS OF TOXIC METALS

We have now reviewed the biological and chemical behavior of a few important metals that affect water quality. There are, of course, many other metals that pose possible environmental threats, and the understanding of these metals as they effect human health can be approached using fundamental principles. To do this it is necessary to understand basic solubility and oxidation–reduction equilibria, for these concepts regulate, for the most part, the movement of metals through environmental systems. We shall now review the fundamental mechanisms and processes that alter the forms of metals in the environment.

A. Solubility and Precipitation

All substances, including metals, have a characteristic solubility that is a function of several parameters. This is a fundamental and important aspect, because the solubility of a toxic metal is directly proportional to its ability to cause environmental damage. Insoluble metals are considered unavailable for biological activity, and hence do not pose severe water quality problems.

Equilibrium is usually established between the solid phase of a metal and its solute phase. In this situation the maximum amount of metal, under the given environmental conditions, will be in solution. When this occurs, it is possible to calculate the solubility product K_s, that is, the product of the concentrations of the ionic species involved in the dissolution. This product is a constant, and it can be used to relate the comparative ability of a given compound to go into solution. For the general reaction

$$M_mA_a \rightarrow mM + aA,$$

the solubility product will be

$$K_s = [M]^m[A]^a.$$

The extent to which any salt dissolves in water can be approximated by its solubility product. As an example, consider the solubility equation for calcium phosphate $(Ca_3(PO_4)_2)$:

$$Ca_3(PO_4)_2 \rightarrow 3Ca + 2PO_4.$$

Hence,

$$K_s = [Ca]^3[PO_4]^2 = 1 \times 10^{-27}$$

at 25°C. If it is known that the solubility of phosphate, for instance, is governed by its precipitation with calcium, then it is possible to calculate the amount of phosphorus present in the solution at equilibrium. That is, by knowing the solubility product and measuring the concentration of calcium in the system, the phosphate concentration expected at equilibrium can be calculated. Table 3.5 shows the solubility product for selected metal com-

TABLE 3.5   HALF REACTIONS AND
SOLUBILITY PRODUCTS FOR SELECTED
METAL COMPOUNDS[a]

EQUILIBRIUM EQUATION	K_s AT 25°C
$MgCo_3 \rightleftharpoons Mg^{2+} + CO_3^{2-}$	4×10^{-5}
$Mg(OH)_2 \rightleftharpoons Mg^{2+} + 2OH^-$	9×10^{-12}
$CaCO_3 \rightleftharpoons Ca^{2+} + CO_3^{2-}$	5×10^{-9}
$Ca(OH)_2 \rightleftharpoons Ca^{2+} + 2OH^-$	8×10^{-6}
$Cu(OH)_2 \rightleftharpoons Cu^{2+} + 2OH^-$	2×10^{-19}
$Zn(OH)_2 \rightleftharpoons Zn^{2+} + 2OH^-$	3×10^{-17}
$Zn(OH)_2 \rightleftharpoons 2H^+ + ZnO_2^{2-}$	1×10^{-29}
$Ni(OH)_2 \rightleftharpoons Ni^{2+} + 2OH^-$	2×10^{-16}
$Cr(OH)_3 \rightleftharpoons Cr^{3+} + 3OH^-$	6×10^{-31}
$Cr(OH)_3 \rightleftharpoons CrO_2^- + H^+ + H_2O$	9×10^{-17}
$Al(OH)_3 \rightleftharpoons Al^{3+} + 3OH^-$	1×10^{-32}
$Al(OH)_3 \rightleftharpoons H^+ + AlO_2^- + H_2O$	4×10^{-13}
$Fe(OH)_3 \rightleftharpoons Fe^{3+} + 3OH^-$	6×10^{-38}
$Fe(OH)_2 \rightleftharpoons Fe^{2+} + 2OH^-$	5×10^{-15}
$Mn(OH)_3 \rightleftharpoons Mn^{3+} + 3OH^-$	1×10^{-36}
$Mn(OH)_2 \rightleftharpoons Mn^{2+} + 2OH^-$	8×10^{-14}
$Ca_3(PO_4)_2 \rightleftharpoons 3Ca^{2+} + 2PO_4^{3-}$	1×10^{-27}
$CaHPO_4 \rightleftharpoons Ca^{2+} + HPO_4^{2-}$	3×10^{-7}
$CaF_2 \rightleftharpoons Ca^{2+} + 2F^-$	3×10^{-11}
$AgCl \rightleftharpoons Ag^+ + Cl^-$	3×10^{-10}
$BaSO_4 \rightleftharpoons Ba^{2+} + SO_4^{2-}$	1×10^{-10}

[a] From "Fundamentals of Chemistry for Environmental Engineering," 3rd ed., by C. Sawyer and P. McCarty, © 1978. Reproduced with permission from McGraw-Hill Book Company.

pounds. Note that the solubilities of the different metals are variable and that solubilites of the same metal complexed with different ions are also variable. For instance, the solid phase of magnesium carbonate has a solubility product of abut 4×10^{-5}, whereas K_s for magnesium complexed with hydroxide is about 9×10^{-12}. In environmental situations, as we shall see later, the use of solubility equilibria to predict resultant metal concentrations must be done with care. Equilibrium is seldom if ever achieved in the environment, and the presence of competitive ions and/or organics in solution can change the equilibrium solubility as calculated using a single K_s value.

In addition to the dissolution of metals, we must also be concerned with precipitation, that is, the formation of the solid phase from the aqueous phase. Here the solubility product is used with a term called the "common ion effect." If an ion added to a salt solution is identical to one of the ions already in the dissolved salt, the solubility of the other ion decreases. Referring again to our calcium–phosphate model and recalling that $K_s = [Ca]^3[PO_4]^2$, it is seen that when calcium is added to the water, for the solubility product to remain constant, the phosphate concentration must decrease. This is a common occurrence in estuarine systems, in which many of the metal salts are chlorides. Once the dissolved metal reaches waters containing higher chloride levels, the metal solubility must decrease because of the common ion effect, and the metal will precipitate such that the solubility product with respect to chloride remains constant.

It has been shown that simple, fundamental chemical equilibria can be applied to make predictions of the solubility of metals of water quality interest. These data must be used with caution in natural systems, however, and only in a qualitative sense. This is because it is difficult to determine which soluble form is governing the solubility of a particular metal. We must now address this question, because it is essential in quantifying the solubility relationships of metals.

The complexation preference of metals is a function of several parameters. Ahrland et al. (1958) divided metals into two categories that reflect their preference for certain ligands in complex formation (see Table 3.6). Note that class A metals complex with flouride in preference to ammonia or cyanide. Sulfide precipitates are not expected, because hydroxide radicals are normally bound by other atoms before sulfur. Class B metals tend to bind ammonia in preference to hydroxide and to form stable chloride complexes. These metals also tend to form insoluble sulfides, which was not the case for the class A metals.

It should be emphasized again that the occurrence of all soluble and insoluble species of a given metal must be evaluated. As shown in Table 3.5, solubility products vary over many orders of magnitude and can lead to great errors in predictions of resultant metal concentrations if the species

TABLE 3.6   CLASSIFICATION OF METALS
WITH RESPECT TO LIGAND PREFERENCE IN
COMPLEX FORMATION[a]

METAL CATIONS	LIGAND PREFERENCE
Class A	
H^+, Li^+, Na^+, K^+,	$F > O > N = Cl > Br >$
Be^{2+}, Mg^{2+}, Ca^{2+}, Sr^{2+},	$I > S$, $OH^- > RCO_2^-$,
Al^{3+}, Sc^{3+}, La^{3+}, Si^{4+},	$CO_3^{2-} > NO_3^-$,
Ti^{4+}, Zr^{4+}, Th^{4+}	$PO_4^{3-} > SO_4^{2-}$
Class B	
Cu^+, Ag^+, Au^+, Te^+,	$S > I > Br > Cl = N >$
Ga^+, Zn^{2+}, Cd^{3+}, Hg^{2+},	$O > F$
Pb^{2+}, Sn^{2+}, Tl^{3+}, Au^{3+},	
In^{3+}, Bi^{3+}	

[a] From Ahrland (1958).

governing solubility is unknown. Although the theoretical concepts involved in constructing phase diagrams for these predictions are beyond the scope of this book, some general classifications can be made. The class A metals in Table 3.6 tend to form hydroxides or oxides as stable precipitates in the pH range of natural waters. This is especially true of the trivalent and tetravalent metal ions. Phosphorus compounds, for example, are readily formed with Al^{3+} and Fe^{3+} but are stable only at low pH values. The carbonate ion as well as the phosphate ion are successful competitors with oxides and hydroxides for many metal ions. Thus, in waters containing an appreciable carbonate concentration, as is common in most natural systems, one can readily find ferrous carbonate, calcium carbonate, and other carbonates in the solid phase.

According to Stumm and Morgan (1970), it is necessary to consider all the inorganic complexes to determine the solubility of metals. In marine and estuarine waters, the chloride ion appears to dominate over hydroxide for Ag^+, Hg^{2+}, Au^{2+}, and probably Cd^{2+} and Pb^{2+}. Thus, the solubility of these particular metals is dominated by the chloride complex.

B. Chelation

The discussion thus far has centered around inorganic complexes of metal ions, and it has been shown that basic solubility relationships can be applied to predict resultant soluble concentrations of metals in aqueous systems.

TABLE 3.7 CONCENTRATION FACTORS OF VARIOUS METALS IN MARINE ORGANISMS[a]

ORGANISM	METAL									
	Cu	Ni	Pb	Co	Zn	Mn	Mg	Ca	Sr	Ba
Seaweed	—	550	—	—	900	—	—	—	23	—
Benthic algae	—	2,000 to 40,000	—	—	—	1,000 to 30,000	—	—	—	—
Plankton	400 to 90,000	<20 to 8000	30 to 12,000	<100 to 16,000	—	—	—	—	—	—
Marine animals	—	3,000 to 70,000	—	—	—	2,000 to 10,000	—	—	—	—
Anchovy	80	—	10,000	—	400	1000	0.1	7	8	20
Yellow fin tuna	200	—	—	—	700	80	0.2	6	7	2
Skipjack tuna	100	50	—	—	500	40	0.3	5	5	3
Sponge	1400	420	—	50	—	—	—	0.07	3.5	—

[a] From Goldberg (1963). © 1963 John Wiley & Sons, Inc.

However, we have yet to consider the phenomenon of chelation, which refers to the formation of organic metal compounds. Most organic materials, including those found in natural systems, are capable of complexing metal ions, with the result that the solubility of metals is higher than predicted by inorganic solubility constants.

Perhaps one of the most important results of the chelation of metals is the uptake of metals by biological organisms. Although little is known about the natural chelation of metals, it is known that the metabolic processes of many organisms chelate metals. The result is that large concentrations of metals can be incorporated into biomass.

Goldberg (1963) has evaluated the concentration factors of various metals in marine organisms. Table 3.7 shows these data, and it can be seen that many metals are concentrated to levels hundreds or thousands of times higher than the ambient concentration in the water. This is owing mainly to chelation, that is, the binding of a trace metal with an organic matrix, and subsequent uptake by the organism.

The effect of chelation of toxic metals is a general increase in the solubility of the metals. However, very little quantitative information is available that relates the extent of organically mediated metal dissolution in natural systems. Water quality predictions must therefore assume that chelation will occur and will be especially pronounced in systems with high amounts of dissolved organic matter. We can therefore expect that large organic inputs will bring more toxic metals into solution. It can be seen that an evaluation of the solubility of metals depends not only on determining the inorganic insoluble species but also on a consideration of dissolved organic complexes. Even so, the most sophisticated evaluation using equilibrium methods can serve as little more than an estimate of the solubility of metal ions.

C. Oxidation – Reduction Equilibria

We shall now consider another parameter that regulates the movement of metals in aqueous systems, that is, the oxidation – reduction potential. Redox reactions are analogous to acid – base reactions except that instead of hydrogen ions being transferred, electrons are transferred. Thus, an element can be oxidized or reduced during a reaction, depending on whether it accepts or donates electrons. Table 3.8 shows the common activity series of metals. The metals are listed in order of increasing activity, that is, the most reactive metals are at the top of the list and the less active metals at the bottom.

Redox reactions can be split into two half reactions, one involving

TABLE 3.8   RELATIVE OXIDATION–
REDUCTION POTENTIAL OF METALS

METAL	COMMENTS
K	React violently, with acids
Ca	
Na	
Mg	React with acids to yield H_2
Al	
Mn	
Zn	
Cr	
Fe	
Cd	
Co	
Ni	
Sn	
Pb	
H	React with O_2 to yield oxides
Cu	
Hg	
Ag	Form oxides indirectly
Pt	
Au	

oxidation or loss of electrons and the other involving reduction or gain of electrons. In computing redox potentials, the potentials of half reactions, which are usually known, are combined to yield the total potential for the system.

The potential for a half reaction is usually referred to as the standard electrode potential E_0. Many tabulations are available that list the voltage or potential required for either the oxidation or reduction of various elements. Table 3.9 shows a list of elements in the oxidized and reduced form with the potential required for the reduction reaction. Here, again, it should be emphasized that E_0 reflects the potential for the half reaction. If two reactions are combined, that is, the oxidation of one material by the reduction of another, we must add the half reaction potentials to obtain the total potential for the reaction. As an example, consider the oxidation of zinc by copper. Referring again to Table 3.9, it is seen that

$$Zn(S) \rightarrow Zn^{2+} + 2e^-, \qquad E_0 = 0.76 \quad V$$

and

$$Cu^{2+} + 2e^- \rightarrow Cu(S), \qquad E_0 = 0.34 \quad V.$$

TABLE 3.9 STANDARD ELECTRODE POTENTIALS FOR SELECTED REACTIONS[a]

REACTION	log K AT 25°C	STANDARD ELECTRODE POTENTIAL (V) AT 25°C
$Na^+ + e^- = Na(s)$	−46.0	−2.71
$Zn^{2+} + 2e^- = Zn(s)$	−26.0	−0.76
$Fe^{2+} + 2e^- = Fe(s)$	−15.0	−0.44
$Co^{2+} + 2e^- = Co(s)$	−9.5	−0.28
$V^{3+} + e^- = V^{2+}$	−8.8	−0.26
$2H^+ + 2e^- = H_2(g)$	0.0	0.00
$S(s) + 2H^+ + 2e^- = H_2S$	0.47	0.14
$Cu^{2+} + e^- = Cu^+$	2.7	0.16
$AgCl(s) + e^- = Ag(s) + Cl^-$	3.7	0.22
$Cu^{2+} + 2e^- = Cu(s)$	12.0	0.34
$Cu^+ + e^- = Cu(s)$	18.0	0.52
$Fe^{3+} + e^- = Fe^{2+}$	13.2	0.77
$Ag^+ + e^- = Ag(s)$	13.5	0.80
$Fe(OH)_3(s) + 3H^+ + e^- = Fe^{2+} + 3H_2O$	18.8	1.06
$IO_3^- + 6H^+ + 5e^- = \frac{1}{2}I_2(s) + 3H_2O$	104.0	1.23
$MnO_2(s) + 4H^+ + 2e^- = Mn^{2+} + 2H_2O$	42.0	1.23
$Cl_2(g) + 2e^- = 2Cl^-$	46.0	1.36
$Co^{3+} + e^- = Co^{2+}$	31.0	1.82

[a] From Stumm and Morgan (1970). © 1970 John Wiley & Sons, Inc.

The total reaction is the sum of the two, that is,

$$Zn(S) + Cu^{2+} \rightarrow Zn^{2+} + Cu(S), \quad E_0 = 0.76 + 0.34 = 1.1 \quad V.$$

This means that a 1.1-V potential must be realized for this reaction to proceed. In general, a positive voltage indicates a spontaneous reaction, or that the reaction will proceed as written. If the voltage had turned out to be negative, e.g., −1.1 V, then the reaction would not be spontaneous or thermodynamically feasible.

1. The Effect of Reactant Concentrations

The calculations in the previous section were based on the concentrations of the ionic reactants and products being equal to unity. If any other concentration is involved, then a change in redox potential will result. The change in potential due to different concentrations of reactants and products is given by the Nernst equation. For the general reaction

$$aA + bB \rightleftarrows cC + dD, \tag{1}$$

which can be broken into the half reactions

$$aA \rightleftarrows cC + xe^-, \tag{2}$$

$$bB + xe^- \rightleftarrows dD, \tag{3}$$

the Nernst equation can be written as

$$E(t) = E_0(t) - \frac{RT}{n} \log \frac{[cC][dD]}{[aA][bB]}. \tag{4}$$

In Eq. (4) the quantities in brackets are the concentrations of the reactants and products, R is the universal gas constant, T the absolute temperature, and n the number of electrons transferred in each half reaction.

The value for $E(t)$ is essentially a measure of the tendency for a reaction to occur. When the net reaction reaches equilibrium, $E(t) = 0$, and the Nernst Equation becomes

$$E_0(t) = \frac{RT}{n} \log \frac{[cC][cD]}{[aA][bB]}, \tag{5}$$

where the terms in brackets are now the *equilibrium* concentrations of reactants and products.

2. pϵ and Oxidation–Reduction

We have seen that the redox potential, or the energy involved in the transfer of one mole of electrons from an oxidant to hydrogen, can be expressed in volts. It is also possible to relate the energy transferred using a dimensionless term, pϵ, which is related to E_0 by

$$p\epsilon = (F/2.3RT)E_0, \tag{6}$$

where F is Faraday's number and the other terms are as defined previously. Conceptually, pϵ can be considered to be the negative logarithm of the electron activity. This is similar to pH, which is the negative logarithm of the relative proton activity. Just as with pH, low values of pϵ reflect high electron activity, and corresponding reducing conditions, whereas large positive values of pϵ, or low electron activity, represent very strong oxidizing conditions. Values for pϵ vary between approximately 14 and -10 in natural systems.

There are other special forms of pϵ defined in the literature, including $p\epsilon^0$, which is the relative electron activity with all other species other than the electrons at unit activity. In addition, the popular symbol $p\epsilon^0(w)$ refers to the relative electron activity of a system when $[H^+]$ and $[OH^-]$ are taken at their activities in neutral water rather than at unity. Values of $p\epsilon^0(w)$ apply to activities of oxidants and reductants at a pH of 7.0, which is more reflective of natural systems. Table 3.10 lists some redox reactions common to natural aquatic systems and the corresponding values for both $p\epsilon^0$ and $p\epsilon^0(w)$.

3. $p\epsilon - pH$ Interactions

It has been shown that the equilibrium phase of certain metals in natural systems is dependent on both pH and $p\epsilon$. It is possible to relate the solubility of metals to pH changes at constant $p\epsilon$ and to calculate the oxidation state of metals at constant pH. A plot of $p\epsilon$ vs. pH could be constructed to reflect the interaction of both oxidation – reduction potential and pH of a metal – ion equilibrium. Although the theoretical equilibrium relationships required to construct $p\epsilon$ – pH diagrams cannot be covered in

TABLE 3.10   OXIDATION–REDUCTION EQUILIBRIA
COMMON TO NATURAL AQUATIC SYSTEMS

REACTION	$p\epsilon^0 (\equiv \log K)$	$p\epsilon^0(w)$
$\frac{1}{4}O_2(g) + H^+(w) + e^- = \frac{1}{2}H_2O$	20.75	13.75
$\frac{1}{5}NO_3^- + \frac{6}{5}H^+(w) + e^- = \frac{1}{10}N_2(g) + \frac{3}{5}H_2O$	21.05	12.65
$\frac{1}{2}MnO_2(s) + \frac{1}{2}HCO_3^-(10^{-3}) + \frac{3}{2}H^+(w) + e^-$	—	8.50
$\quad = \frac{1}{2}MnCO_3(s) + \frac{1}{2}H_2O$		
$\frac{1}{2}NO_3^- + H^+(w) + e^- = \frac{1}{2}NO_2^- + \frac{1}{2}H_2O$	14.15	7.15
$\frac{1}{8}NO_3^- + \frac{5}{4}H^+(w) + e^- = \frac{1}{8}NH_4^+ + \frac{3}{8}H_2O$	14.90	6.15
$\frac{1}{6}NO_2^- + \frac{4}{3}H^+(w) + e^- = \frac{1}{6}NH_4^+ + \frac{1}{3}H_2O$	15.14	5.82
$\frac{1}{2}CH_3OH + H^+(w) + e^- = \frac{1}{2}CH_4(g) + \frac{1}{2}H_2O$	9.88	2.88
$\frac{1}{4}CH_2O + H^+(w) + e^- = \frac{1}{4}CH_4(g) + \frac{1}{4}H_2O$	6.94	−0.06
$FeOOH(s) + HCO_3^-(10^{-3}) + 2H^+(w) + e$	—	−1.67
$\quad = FeCO_3(s) + 2H_2O$		
$\frac{1}{2}CH_2O + H^+(w) + e^- = \frac{1}{2}CH_3OH$	3.99	−3.01
$\frac{1}{6}SO_4^{2-} + \frac{4}{3}H^+(w) + e^- = \frac{1}{6}S(s) + \frac{2}{3}H_2O$	6.03	−3.30
$\frac{1}{8}SO_4^{2-} + \frac{5}{4}H^+(w) + e^- = \frac{1}{8}H_2S(g) + \frac{1}{2}H_2O$	5.75	−3.50
$\frac{1}{8}SO_4^{2-} + \frac{9}{8}H^+(w) + e^- = \frac{1}{8}HS^- + \frac{1}{2}H_2O$	4.13	−3.75
$\frac{1}{2}S(s) + H^+(w) + e^- = \frac{1}{2}H_2S(g)$	2.89	−4.11
$\frac{1}{8}CO_2(g) + H^+(w) + e^- = \frac{1}{8}CH_4(g) + \frac{1}{4}H_2O$	2.87	−4.13
$\frac{1}{6}N_2(g) + \frac{4}{3}H^+(w) + e^- = \frac{1}{3}NH_4^+$	4.68	−4.68
$\frac{1}{2}(NADP^+) + \frac{1}{2}H^+(w) + e^- = \frac{1}{2}(NADPH)$	−2.00	−5.50
$H^+(w) + e^- = \frac{1}{2}H_2(g)$	0.00	−7.00
Oxidized ferredoxin $+ e^- = $ reduced ferredoxin	−7.10	−7.10
$\frac{1}{4}CO_2(g) + H^+(w) + e^- = \frac{1}{24}$(glucose) $+ \frac{1}{4}H_2O$	−0.20	−7.20
$\frac{1}{2}HCOO^- + \frac{3}{2}H^+(w) + e^- = \frac{1}{2}CH_2O + \frac{1}{2}H_2O$	2.82	−7.68
$\frac{1}{4}CO_2(g) + H^+(w) + e^- = \frac{1}{4}CH_2O + \frac{1}{4}H_2O$	−1.20	−8.20
$\frac{1}{2}CO_2(g) + \frac{1}{2}H^+(w) + e^- = \frac{1}{2}HCOO^-$	−4.83	−8.73

[a] From Stumm and Morgan (1970). © 1970 John Wiley & Sons, Inc.

this book, it is informative to review some examples. The reader is referred to Stumm and Morgan (1970) for a detailed treatment of $p\epsilon$–pH diagrams.

As noted earlier, the form of lead complex found in natural systems depends in part on the anions present. It is possible for lead to be complexed as an oxide, hydroxide, carbonate, sulfate, etc., hence the $p\epsilon$–pH diagram constructed will depend on the insoluble phases present in the system. Figure 3.6 shows a typical $p\epsilon$–pH diagram for a Pb–CO_2–H_2O system. It is shown that, for normal pH values and $p\epsilon$ situations common to natural water, lead carbonate should be the predominant species at equilibrium. At lower pH values Pb^{2+} should be the thermodynamically favored species, and at lower redox potentials the formation of elemental lead should be favored.

Hem (1963) has constructed a similar diagram for an Fe–H_2O system (see Fig. 3.7). Here, the reader should recall the relation of E_0 to $p\epsilon$ defined in Eq. (6). In similar aqueous systems iron behaves in a similar manner to

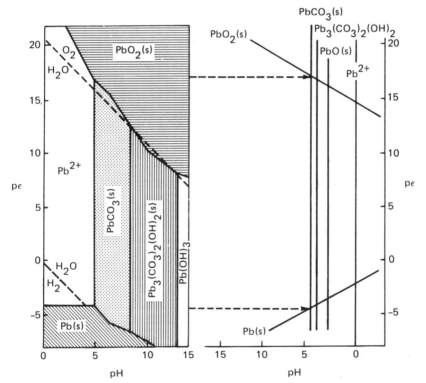

FIGURE 3.6 $p\epsilon$ VS. pH FOR THE Pb–CO_2–H_2O SYSTEM. (From Stumm and Morgan, 1970. © John Wiley & Sons, Inc.)

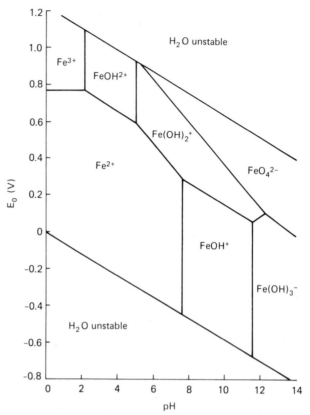

FIGURE 3.7 E_0 VS. pH FOR THE Fe–H_2O SYSTEM.   (From Hem, 1963.)

lead, and one could make predictions as to the species of iron present in water at equilibrium as a function of E_0 and pH.

The preceding examples summarize the physical and chemical factors that affect the pε–pH equilibrium for trace metals in natural systems. Included in this is an understanding that the solid phases governing solubility must also be known. In addition, the redox potential of the water system must be understood as it affects the solubility of certain metals. It will now be shown that certain elements are much more readily affected by changes in redox potential than others. The elements that are capable of being oxidized or reduced in natural systems generally pose the greatest water quality problems. We shall now focus on these particular elements and show how pε–pH relationships can be used to predict potential effects on water quality.

4. Oxidation and Reduction in Natural Systems

Oxidation–reduction potentials are useful measures of the oxidation state to which a particular element moves in a natural system. Relatively few elements in natural systems are oxidized or reduced by changes in the redox potential of the system. In general, carbon, nitrogen, oxygen, sulfur, iron, and manganese are the predominant elements involved in natural redox processes. It is interesting to note that these are also the main elements involved in biological activity.

The major complication associated with using redox parameters for predicting water quality effects is the inherent dynamic state of natural systems, that is, equilibrium is seldom achieved. This is owing mostly to biological activity, in particular photosynthesis. It is quite possible to find reduced species of elements in a system that is in an oxygenated state (for example, with a $p\epsilon$ of 13.6 and a pH of 7), even though equilibrium calculations predict that all of the elements should be in their most oxidized state. Recall that photosynthesis is the process of fixing oxidized forms of nutrients into a reduced or cellular state. With this in mind, it can be seen that redox equilibrium methods are capable of predicting only the thermodynamic state of a system, or the equilibrium state toward which a natural system should be moving.

Redox reactions are generally much slower than pH equilibrium reactions, unless biological mediation occurs. Therefore, the environmental movement of most elements of interest in terms of water quality is biologically mediated. Table 3.11 lists the common microbially mediated oxidation–reduction reactions in natural systems. Each of the half reactions has a corresponding value of $p\epsilon^0(w)$, and it is at this oxidation–reduction level that the particular process operates. Given this information, it is possible to estimate qualitatively the fate of certain elements by observing the oxidation–reduction potential of a given system. For instance, nitrate (NO_3^-) would be reduced to nitrogen gas at a $p\epsilon^0(w)$ of 12.6, and sulfate (SO_4^{2-}) would be reduced to HS^- at a $p\epsilon^0(w)$ of -3.75. Estimates can thus be made to determine the oxygen level that should be maintained in a system to assure the occurrence of the oxidized species of these elements. This is of great importance in view of the frequent contamination of water by reduced species of both sulfur and nitrogen.

The measurement of E_0 in a system is qualitatively very simple and requires only an inert metal electrode (usually platinum) in conjunction with a reference electrode. These measurements, though seemingly simple, do not, however, reflect the actual oxidation–reduction potential. This is owing again to the dynamic situation in natural systems, in which several ionic species can affect the measurement simultaneously. Hence, as

TABLE 3.11 COMMON MICROBIALLY MEDIATED OXIDATION–REDUCTIONS[a]

OXIDATION	$pe^0(w)$ $(= -\log K (w))$	REDUCTION	$pe^0(w)$ $(= \log K)$
$\frac{1}{4}CH_2O + \frac{1}{4}H_2O = \frac{1}{4}CO_2(g) + H^+(W) + e^-$	−8.20	$\frac{1}{4}O_2(g) + H^+(W) + e^- = \frac{1}{2}H_2O$	13.75
$\frac{1}{2}HCOO^- = \frac{1}{2}CO_2(g) + \frac{1}{2}H^+(W) + e^-$	−8.73	$\frac{1}{5}NO_3^- + \frac{6}{5}H^+(W) + e^-$ $= \frac{1}{10}N_2(g) + \frac{3}{5}H_2O$	12.65
$\frac{1}{4}CH_2O + \frac{1}{2}H_2O$ $= \frac{1}{4}HCOO^- + \frac{5}{4}H^+(W) + e^-$	−7.68	$\frac{1}{8}NO_3^- + \frac{5}{4}H^+(W) + e^-$ $= \frac{1}{8}NH_4^+ + \frac{3}{8}H_2O$	6.15
$\frac{1}{2}CH_3OH = \frac{1}{2}CH_2O + H^+(W) + e^-$	−3.01	$\frac{1}{2}CH_2O + H^+(W) + e^- = \frac{1}{2}CH_3OH$	−3.01
$\frac{1}{2}CH_4(g) + \frac{1}{2}H_2O$ $= \frac{1}{2}CH_3OH + H^+(W) + e^-$	2.88	$\frac{1}{8}SO_4^{2-} + \frac{9}{8}H^+(W) + e^- = \frac{1}{8}HS^- + \frac{1}{2}H_2O$	−3.75
$\frac{1}{8}HS^- + \frac{1}{2}H_2O$ $= \frac{1}{8}SO_4^{2-} + \frac{9}{8}H^+(W) + e^-$	−3.75	$\frac{1}{8}CO_2(g) + H^+(W) + e^- = \frac{1}{8}CH_4(g) + \frac{1}{4}H_2O$	−4.13
$\frac{1}{8}NH_4^+ + \frac{3}{8}H_2O$ $= \frac{1}{8}NO_3^- + \frac{5}{4}H^+(W) + e^-$	6.16	$\frac{1}{6}N_2 + \frac{4}{3}H^+(W) + e^- = \frac{1}{3}NH_4$	−4.68
$FeCO_3(s) + 2H_2O$ $= FeOOH(s) + HCO_3^-(10^{-3})$ $- 2H^+(W) + e^-$	−1.67		
$\frac{1}{2}MnCO_3(s) + \frac{3}{2}H_2O$ $= \frac{1}{2}MnO_2(s) + \frac{1}{2}HCO_3^-(10^{-3})$ $+ \frac{3}{2}H^+(W) + e^-$	−8.5		

[a] From Stumm and Morgan, 1970. © 1970 John Wiley & Sons, Inc.

Stumm (1966) points out, aerobic systems tend to have large positive pϵ values and anaerobic systems tend to have negative pϵ values. Quantification of values between these extremes, however, is very difficult. It is perhaps a sounder practice to calculate the concentrations of the oxidized and reduced forms of a redox pair and then calculate the oxidation–reduction potential.

It has been shown how physical and chemical processes alter the oxidation and solubility states of toxic metals. The calculations presented in Section III.A show how equilibrium models can be used to describe the stability of certain materials in aqueous systems. Although most natural systems are dynamic in nature, the equilibrium system should be determined so that rational decisions concerning water quality can be made. We shall now see how certain elements undergo biomagnication and investigate the factors governing the biological movement of toxic metals in the environment.

IV. SETTING WATER QUALITY STANDARDS FOR TOXIC METALS

Toxic elements in the environment, especially metals, have received considerable attention in recent years. However, much of the work related to the environmental effects of metals has dealt with direct toxicity to biological systems. McKee and Wolf (1971) compiled an exhaustive list of elements, including most trace metals, and their reported toxicities to biological organisms. The reader is referred to their report for data on particular metals.

A. Measurements of Toxicity

Toxicity data are reported on various base levels; the common methods of expressing toxicity data are as follows:

(1)   MLD, the minimum lethal dose required to kill one or more of a test species;

(2) TL_m, the tolerance limit median, which designates the concentration required to kill 50% of the test organisms;

(3) LD_{50}, the median lethal dose required to kill 50% of the test species.

The values for MLD and TL_m are exposure dependent, that is, one usually considers the time of exposure to the element when reporting these values. On the other hand, the LD_{50} values are usually based on single injections or feedings, so that time of exposure is not of importance. In general, acute toxicity or direct toxicity is referred to as a lethal effect that occurs within a period of 96 h after injection. As would be expected, however, many elements do not cause mortality within 96 h or even within periods of weeks or months. They do, however, cause subtle changes in appetite, disorders of the nervous system, malfunctioning of reproduction, and other symptoms. These and similar effects are referred to as chronic toxicity.

B. Interpreting Toxicity Data

The preceding discussion indicates that setting water quality standards on toxic metal concentrations is a very complicated task. It is generally not clear what the optimum allowable level should be in water systems. For example, it is probably not sufficient to use LD_{50} or MLD data for certain species of fish as a criteria for the discharge of heavy metals, because it is possible that the fish could suffer from chronic effects that are not identified by using LD_{50} data.

The problem then arises, how are the standards or criteria for water quality protection set? Attempts have been made to use fractions of LD_{50} or MLD values as safe limits. Although this would include a large factor of safety based on known toxicity studies, it still cannot account for chronic effects or interactions with other phases of the environment. The Environmental Protection Agency's drinking water standards are based primarily on lethal dose studies. Even for cases in which heavy metals are limited to low concentrations, good medical evidence on actual chronic effects from some of the metals is still not available. It is also well known that many modes of metal uptake other than water consumption are common in humans. These other inputs must be added to the drinking water input, and therefore the values used as safe limits for drinking water must be kept at an extremely conservative level.

It should be noted that most of the bioassay toxicity studies reported by McKee and Wolf (1971) were performed on multicellular aquatic organisms. This may not be the best type of organism to test, because most multicellular organisms contain complex homeostatic systems that can adjust for many environmental stresses. Many of these higher-order animals are capable of adjusting to heavy metal concentrations higher than the values that would be lethal to other components of the system. Several workers, including Carter and Cameron (1973), have proposed the use of single-celled animals in bioassay experiments. In particular, they have used the ciliated protozoan *Tetrahymena pyriformis* as a test organism. Their work has shown that when protozoa are introduced to a variety of toxic metals, cadmium is the most toxic metal, followed in order by mercury, cobalt, zinc, and lead. This ranking of toxicity is similar to the data obtained in the fish studies, except for the ranking of lead. They concluded that single-celled protozoa are a more sensitive indicator of heavy metal toxicity than are multicellular animals such as fish.

C. Relative Toxicities of Heavy Metals

When considering the possible impact of trace metals on water quality, it is necessary to evaluate the biological cycles of these elements. Toxicity data are important and are most useful for setting standards. There must be an understanding, however, of which metals are biologically active and hence pose the threat of moving into higher trophic levels. Wood (1974) classified heavy metals into three categories:

(1)   noncritical,
(2)   toxic and readily accessible,
(3)   toxic but very insoluble or rare.

Using this classification it is possible to evaluate the relative environmental damage possible from contamination by various metals. Wood considered various physical and chemical processes and then evaluated biological mobilization to derive his classifications. As an example, the movement of mercury can be compared with that of cadmium. In general, the low vapor pressure of mercury accounts for its dispersal throughout the world, and the alkylmercury compounds, as shown previously, are quite stable in the environment, which leads to biomagnification. Cadmium, on the other hand, does not form stable alkyl-metals in natural systems, so the ability of cadmium to go into solution depends only on the solubility of the cadmium salts. Also, as Wood points out, microbial metabolism of elements is a good indication of their general biological activity, for if microorganisms are unable to mobilize an element, then higher organisms probably cannot do so either.

Mercury in its elemental form is extremely volatile, and hence it can be lost from aqueous systems. On the other hand, it *can* exist as stable alkylmethylmercury or dimethylmercury. Many microorganisms are capable of detoxifying methylmercury by reducing it to elemental mercury and methane.   This process is common in water systems, but it has also been shown that many chemical and biological processes are operative in producing elemental mercury, methylmercury, and other mercury salts.   The measurement of any particular form of mercury in the environment, then, does not yield meaningful results.   It is important to determine the total mercury in a given water system so that possible alkylation rates can be applied.   Figure 3.8 shows a biological cycle for mercury.   It can be seen that many interactions are operative, and many specific rates must be known to fully define the cycle.

Using the information we have compiled on the behavior of mercury and its various phases in environmental systems, it is now possible to make predictions of the ability of other heavy metals to be alkylated in the environment.   For example, Wood (1974) predicts that tin, paladium,

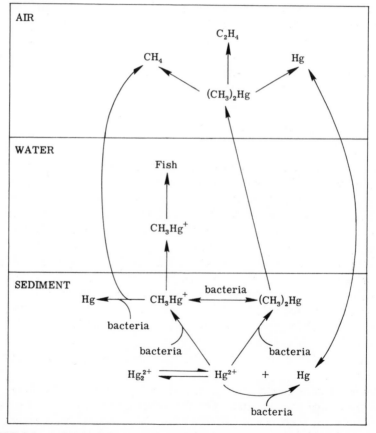

FIGURE 3.8   A BIOLOGICAL CYCLE FOR MERCURY.   (From Wood, 1974. ©
1974 by the American Association for the Advancement of Science.)

platinum, gold, and thalium can be methylated in the environment but that lead, cadmium, and zinc cannot. This is because the alkyl-metals of cadmium, zinc, and lead are not stable in aqueous systems. These evaluations are important for predicting water quality effects, because several of the metals listed are capable of being methylated and are presently used in various industrial processes.

The following list shows Wood's (1974) classification of elements according to their toxicity in aqueous systems.

(1)   Noncritical: Na, K, Mg, Ca, H, O, N, C, P, Fe, S, Cl, Br, F, Li, Rb, Sr, Al, Si.

(2)   Very toxic and relatively accessible: Be, Co, Ni, Cu, Zn, Sn, As, Se, Te, Pd, Ag, Cd, Pt, Av, Hg, Te, Pb, Sb, Bi.

(3)   Toxic but very insoluble or very rare: Ti, Hf, Zr, W, Nb, Ta, Re, Ga, La, Os, Rh, Ir, Ru, Ba.

The elements in (2), that is, those that are toxic and relatively accessible, are obviously the elements of greatest concern for water quality. It should be noted that the metals copper, zinc, lead, cadmium, and mercury are included in this category.

D.   The Validity of Threshold Limits

It is useful at this point to introduce the "no-threshold limit" concept of chemicals in the environment. It can be argued that the occurrence of even one atom of a toxic material entering a cell will cause cellular damage. The basis of this argument is that any time a molecule enters a biological system, a chemical–biological interaction results. It is not clear, however, whether these reactions are always damaging to the cell. In any case, proponents of this concept feel that the occurrence of any amount of toxic metal in the environment is capable of causing great ecological damage, and thus there is no safe level for the discharge of these materials.

Dinman (1972) contends that the concept of no-threshold limit is not valid. That is, the theory does not conform to normal stochastic processes. He contends that there are many examples showing that cells accumulate "toxic elements" and still do not appear to be harmed in any manner. He also contends that there are many interfering substances in a cell that offer a stochastic alternative for a potentially harmful element. Thus, the interaction of a damaging element with an active site in any cell is statistically predictable. There should therefore be some lower limit at which these elements can occur in nature without causing biological damage. However, although these considerations are important from a water quality perspective, there appears to be little solid evidence for either case.

V.   CONCLUSION

In summary, the occurrence of toxic metals in the environment and their potential for biomagnification are functions of several processes. The oxidation or reduction and pH status of the system, solubility characteristics, and, perhaps most importantly, microbial mediation all affect the mobility of metals. When a metal goes into solution in certain environmental situations, it can present an environmental problem. In particular, it was shown that metals that form soluble alkyls in aqueous systems cause the greatest water quality problems.

Owing to the tremendous number of interactions related to the movement of metals in the environment, concentration predictions based on the measurement of any single form of a metal appear to be useless. Thus, the monitoring of heavy metals in the environment should emphasize measuring total concentrations of each metal and not limit quantification to one particular form.

Monitoring toxic metals and performing bioassays for toxicity also need a great deal of refinement. We have seen that lethal-dose studies probably do not reflect the chronic effects of metals at low concentrations in the environment. Most waste treatment processes make no allowance for the removal of toxic metals, and other than the small fraction that is removed by the reduction of organic matter, very little progress has been made in treating toxic metals. Therefore, often the most toxic elements in a waste stream are not affected by the normal waste treatment processes in current use. It is also important that a uniform and appropriate test organism be selected so that chronic effects from low-level metal toxicity can be determined. Assuming that the concept of "no-threshold limit" for metals is not valid, then a reasonable limit must be set for the ultimate safety of aquatic systems. This can only be accomplished when reliable data are available that can relate the actual effects of toxic metals on natural systems.

REFERENCES

Arhland, S., Clatt, S. J., and Davies, W. R. (1958). *Q. Rev. (London)* **12,** 165.
Bittell, J. E., and Miller, R. J. (1974). Lead, cadmium, and calcium selectivity coefficients on montmorillonite, illite and kaolinite. *J. Environ. Qual.* **3,** 250–253.
Bowen, H. J. M. (1960). "Trace Elements in Biochemistry." Academic Press, New York.
Cannon (1969). *In* "Trace Substances in Environmental Health" (D. D. Hemphill, ed.). University of Missouri Press, Columbia.

Carter, J. W., and Cameron, I. L. (1973). Toxicity bioassay of heavy metals in water using *Tetrahymena phriformis.* *Water Res.* 7, 951–961.

Dinman, B. D. (1972). Non-concept of no-threshold chemicals in the environment. *Science* 175, 495–497.

Fagerström, T., and Jernelöv, A. (1972). Some aspects of the quantitative ecology of mercury. *Water Res.* 6, 1193–1212.

Goldberg, E. (1963). Minor elements in seawater. *In* "The Sea" (M. N. Hill, ed), Ch. 1. Wiley (Interscience), New York.

Goldwater, L. J., and Clarkson, T. W. (1972). Mercury. *In* "Metallic Contaminants and Human Health" (Douglas Lee, ed.). Academic Press, New York.

Hahne, H. C. H., and Kroontje, W. (1973). Significance of pH and chloride concentration on behavior of heavy metal pollutants: Mercury (II), Cadmium (II), Zinc (II), and Lead (II). *J. Environ. Qual.* 2, 444–450.

Hem, J. D. (1963). U. S. Geological Survey Water Supply Paper 1667-A.

Kopp, C. J. (1969). "Trace Substances in Environmental Health" (D. D. Hemphill, ed.). University of Missouri Press, Columbia.

McKee, J. E., and Wolf, H. W. (1971). Water Quality Criteria Publication 3-A. California State Water Resources Control Board, Sacramento, California.

Sawyer, C., and McCarty, P. (1978). "Fundamentals of Chemistry for Environmental Engineering," 3rd ed. McGraw-Hill, New York.

Skaar, H., Ophus, E., and Gullvåg, B. M. (1973). Lead accumulation within nuclei of moss leaf cells. *Nature* 241, 215–216.

Spangler, W. J., Spigarelli, J. L., Rose, J. M., Flippin, R. S., and Miller, H. H. (1973). Degradation of methylmercury by bacteria isolated from environmental samples. *J. Appl. Microbiol.* 25, 488–493.

Stumm, W. (1966). Redox potential as an environmental parameter; conceptual significance and operational limitation. *In* Third International Conference on Water Pollution Research, Munich, Germany. Water Pollution Control Federation, Washington, D. C.

Stumm, W., and Morgan, J. (1970). "Aquatic Chemistry." Wiley, New York.

Symposium on Environmental Lead Contamination (1966). U. S. Public Health Service Publication No. 1440.

Valiela, I., Banus, M. D., and Teal, J. M. (1974). of salt marsh bivalves to enrichment with metal containing sewage sludge and retention of lead, zinc and cadmium by marsh sediments. *J. Environ. Pollut.* 7, 149–157.

Wood, J. M. (1974). Biological cycles for toxic elements in the environment. *Science* 183, 1049–1052.

Chapter 4

Refractory Organic Compounds and Water Quality

I. INTRODUCTION

In this chapter we shall discuss the fate and reactivity of organic matter that is considered to be refractory. The term "refractory" refers to those organic compounds that resist degradation and remain in the environment for long periods of time. As is the case with certain metals, these refractory compounds can enter food webs and accumulate such that organisms at the higher levels of the trophic web exhibit high concentrations in their bodies. One of the main reasons for the bioaccumulation of organics is their relative solubility in fats; thus, certain animals can develop concentrations of organics 1000–10,000 times in excess of the concentration in the ambient water.

The classification of the organic compounds discussed in this chapter is a Herculean task because most of them are synthetic and have been introduced into natural systems by human activity. Included in the general list of refractory organics are pesticides, detergents, and petroleum hydrocarbons. There are also many chemicals used in industrial processes, such as polychlorinated biphenols, that find their way into natural water systems. These refractory organics are then subject to certain biological and chemical processes that can remove them from natural systems. Chapter 3 explained that trace metals in natural systems behave generally according to their theoretical solubility equilibria and oxidation–reduction potentials. The persistence of refractory organics in natural systems depends on different parameters. It is difficult to predict the behavior of any group of organics, because a small molecular change can result in a large change in a compound's resistance to natural degradation. We shall use certain model compounds in this chapter and explain the complexities involved in their degradation processes. It should be remembered, however, that very few general rules can be applied to the breakdown of refractory organics in natural systems.

II. HEALTH EFFECTS

The literature abounds with evidence that indicts many pesticides and other refractory organics as the principle agents responsible for the decline of certain animal species. It has been well documented, for instance, that compounds such as the pesticide DDT build up through the levels of the food chain, with the higher carnivores accumulating high concentrations in their bodies. The compound DDT, which we shall discuss in detail in Section III.B, has been blamed for the decimation of populations of herring gulls, brown pelicans, and cormorants. Primary among the documented effects is the thinning of egg shells. If the eggs of these predator birds contain high DDT concentrations, they can break prematurely, resulting in the death of the embryo. It has also been reported that, when concentrations of DDT reached 5 ppm in fish egg yolks, a 100% mortality rate resulted. Sea lions off the California coast exhibit premature birth when the DDT concentration in the mother's blubber reaches a level of approximately 1000 ppm.

The ultimate threat of refractory organics from a water quality perspective is their possible effects on human health. Very few data on a long-term basis are available for any compound except DDT, and there have been studies of the DDT level in humans for only the last 10–15 yr. Figure 4.1 shows the occurrence of DDT in meat, fish, poultry, dairy products, and humans. The concentration of DDT is quite high in meat, fish, and poultry but not quite so high in dairy products. It is also interesting to note the downward trend of DDT levels in both food and human tissues, which is owing to the fact that the use of DDT was essentially prohibited in 1972.

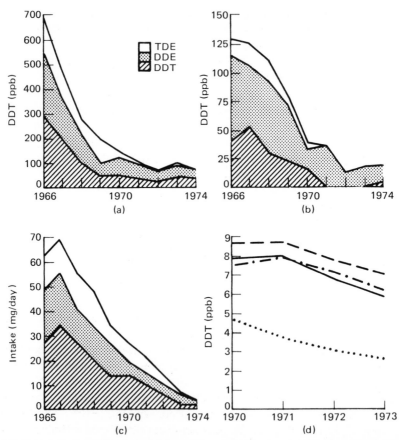

FIGURE 4.1　ESTIMATED HUMAN DIETARY INTAKE AND RESIDUES OF DDT FOR THE FISCAL YEARS 1965–1974: (a) meat, fish, and poultry; (b) dairy products; (c) estimated dietary intake; (d) total DDT in human fatty tissues by age (· · · , ages 0–14; — · —, ages 15–49; ——, ages 50 and higher; ——, total population). (From Council of Environmental Quality, 1981.)

It is difficult to determine exactly the human health hazard of refractory organics in water systems. There have been no recorded deaths attributed to high levels of either pesticides or other refractory organics in natural systems. This, of course, does not include high doses received from industrial accidents. It is known, however, that the occurrence of high levels of organics such as DDT in human fat can aggravate other diseases. When an individual is sick, the fat reserves are burned by the body during the illness. If this fat contains high levels of foreign organics, such as DDT or polychlorinated biphenols, the disease can be aggravated. Thus, many diseases that would not otherwise have been severe can become acute owing to the presence of these compounds in body fat.

III. PESTICIDES AND WATER QUALITY

Pesticides vary with respect to their specificity for target organisms, some being general killers and others being very target specific. Pesticides also vary with respect to persistence in the environment, i.e., some can be degraded very easily and others remain unaltered for long periods of time. We shall examine certain types of pesticides and evaluate the water quality problems that result from their use.

A. Uses of Pesticides

Pesticides include those agents that kill plants, insects, worms, fungi, etc., and are referred to as herbicides, insecticides, nematocides, fungicides, etc. There are thousands of chemicals used as pesticides in the United States, and approximately 50% of all pesticides are used for farming. There are both organic and inorganic pesticides in use, but we shall only consider those of the organic type here. The inorganic forms include biocides such as lead, arsenic, and mercury and can also cause water quality problems. Because the behavior of metals in the environment was discussed in Chapter 3, this chapter will only deal with the organic pesticides.

The three general classes of pesticides that will be discussed in this chapter are chlorinated hydrocarbons, organophosphorus compounds, and a group of general pesticides that includes herbicides, fungicides, and nematocides. This is a general classification, but it is quite useful for predicting water quality effects resulting from the use of these compounds. Because the behavior of pesticides is so variable in natural systems, we shall discuss model compounds to give the reader a feel for the behavior of pesticides in natural systems. Subsequently, the environmental parameters with the most influence on pesticide degradation rates will also be discussed. The breakdown products in many cases can be as toxic or more toxic than the original pesticide. We are therefore not dealing with just a large number of potentially toxic compounds termed pesticides but also the many metabolites that result from their use.

TABLE 4.1 COMMON AND CHEMICAL NAMES OF PESTICIDES AND RELATED COMPOUNDS

COMMON NAME	CHEMICAL NAME
Abate	O,O,O',O'-Tetramethyl-O,O'-thiodi-p-phenylene phosphorothioate
Aldrin	1,2,3,4,10,10-Hexachloro-1,4,4a,5,8,8a-hexahydro-1,4-$endo,exo$-5,8-dimethanonaphthalene
Amizol	2-Amino-1,2,4-triazole
Atrazine	2-Chloro-4-ethylamino-6-isopropylamino-5-trazine
Azinphos-methyl	See guthion
Baytex (fenthion)	O,O-Dimethyl-O-[4-(methylthio)-m-tolyl] phosphorothioate
Bayoxon	O,O-Dimethyl-O-[4-(methylthio)-m-tolyl) phosphate
BHC	Mixture of stereoisomeric 1,2,3,4,5,6-hexachlorocyclohexanes
Chlordane	2,3,4,5,6,7,8,8-Octachloro-2,3,3a,4,7,7a-hexahydro-7-methanoindene
CIPC	Isopropyl N-(3-chlorophenyl)carbamate
Co-Ral	O,O-Diethyl O-3-chloro-4-methyl-2-oxo-2H-1-benzopyran-7-yl phosphorothioate
2,4-D	2,4-Dichlorophenoxy acetic acid
Dalapon	2,2-Dichloropropionic acid
2,4-DB	2,4-Dichlorophenoxy butyric acid
2,4-D esters	Butyl, butoxethanol, isooctyl, isopropyl, propylene glycol butyl ether
DBP	p,p'-Dichlorobenzophenone
DDD	1,1-Dichloro-2,2-bis(p-chlorophenyl)ethane
DDDE	1-Chloro-2,2-bis(p-chlorophenyl)ethylene
DDE	1,1-Dichloro-2,2-bis(p-chlorophenyl)ethylene
DDT	1,1,1-Trichloro-2,2-bis(p-chlorophenyl)ethane
DDVP	O,O-Dimethyl-2,2-dichlorovinyl phosphate
Delnav	2,3-p-Dioxanedithiol-bis(O,O-diethyl phosphorodithioate)
Demeton-O	O,O-Diethyl O-ethyl-2-thioethyl phosphorothioate
Demeton-S	O,O-Diethyl S-ethyl-2-thioethyl phosphorothiolate
Diazinon	O,O-Diethyl-O-(2-isopropyl-4-methyl pyrimidinyl) phosphorothioate
Diazoxon	O,O-Diethyl-O-(2-isopropyl-4-methyl-6-pyrimidyl) phosphate
Dibrom	1,2-Dibromo-2,2-dichloroethyl dimethylphosphate
Dieldrin	1,2,3,4,10,10-Hexachloro-exo-6,7-epoxy-1,4,4a,5,6,7,8,8a-octahydro-1,4-$endo,exo$-5,8-dimenthanonaphthalene
Dimethoate	O,O-Dimethyl S-(N-methylacetamide) phosphorodithioate
Dipterex	O,O-Dimethyl-trichloro-hydroxyethyl phosphonate
Diquat	1,1'-Ethylene-2,2'-bipyridylium dibromide
Di-syston (disulfoton)	O,O-Diethyl S-(2-ethylthioethyl) phosphorodithioate
Diuron	3-(3,4-Dichlorophenyl)-1,1-dimethylurea
DNBP	2-sec-Butyl-4,6-dinitrophenol
DNC	Dinitrocresol
Dursban	3,5,6-Trichloro-2-pyridylphosphorothioate
Endothal	2,3-Dicarboxylic acid, 7-oxobicyclo(2,2,1)heptane
Endrin	1,2,3,4,10,10-Hexachloro-6,7-epoxy-1,4,4a,5,6,7,8,8a-octahydro-1,4,5,8-$endo,endo$-dimethanonaphthalene

TABLE 4.1 *(Continued)*

COMMON NAME	CHEMICAL NAME
EPN	*O*-Ethyl *O*-*p*-nitrophenyl phenyl phosphonothioate
Ethion	*O,O,O',O'*-Tetraethyl *S,S'*-methylene-bis-phosphonothioate
Fenac	2,3,6-Trichlorophenyl acetic acid
Fenchlorophos	See ronnel
Fenuron	3-Phenyl-1,1-dimethylurea
FW-152	4,4'-Dichloro-α-dichloromethylbenzhydrol
Guthion	S-(3,4-Dihydro-4-oxo-1,2,3-benzotriazin-3-ylmethyl)*O,O*-dimethyl phosphorodithioate
Heptachlor	1,4,5,6,7,8,8-Heptachloro-3*a*,4,5,5*a*-tetrahydro-4,7-*endo*-methanoindene
Heptachlorepoxide	1,4,5,6,7,8,8-Heptachloro-2,3-epoxy-2,3,3*a*,4,7,7*a*-Hexahydro-4,7-methanoindene
IMHP	2-Isopropyl-4-methyl-6 hydroxypyrimidine
IPC	Isopropyl *N*-phenylcarbamate
Isodrin	1,2,3,4,10,10-Hexachloro-1,4,4*a*,5,8,8*a*-hexahydro-1,4,5,8-*endo,endo*-dimethanonaphthalene
Kelthane	4,4'-Dichloro-α-(trichloromethyl)benzhydrol
Lindane	γ-isomer of 1,2,3,4,5,6-hexachlorocyclohexane
Maloxon	*O,O*-Dimethyl *S*-(1,2-dicarbethoxyethyl) phosphorothioate
Malathion	*O,O*-Dimethyl *S*-(1,2-dicarbethoxyethyl) phosphorodithioate
MCPA	4-Chloro-2-methyl phenoxyacetic acid
MCPB	α-(4-Chloro-2-methylphenoxy)butyric acid
Mecarbam	S-((Ethoxycarbonyl)methylcarbamoyl)methyl *O,O*-diethyl phosphordithioate
Methoxychlor	1,1,1-Trichloro-2,2-bis(*p*-methoxyphenyl)ethane
Methyl parathion	*O,O*-Dimethyl *O*-*p*-nitrophenyl phosphorothioate
Methyl systox	See demeton-O and -S, substitute methyl for ethyl
Mevinphos	See phosdrin
MMTP	3-Methyl-4-methylthiophenol
Morphothion	*O,O*-Dimethyl *S*-(morpholinocarbonylmethyl) phosphorothioate
Nankor	See ronnel
Nemacide	*O*-(2,4-Dichlorophenyl)*O,O*-diethyl phosphorothioate
OMPA	Octamethylpyrophosphoramide
Paraoxon	Diethyl *p*-nitrophenyl phosphate
Paraquat	1,1'-Dimethyl-4,4'-dipyridilium
Parathion	*O,O*-Dimethyl-*O*-*p*-nitrophenyl phosphorothioate
PCP	Pentachlorophenol
Perthane	2,2-Dichloro-1,1-bis(*p*-ethylphenyl)ethane
Phenkapton	*O,O*-Diethyl *S*-(2,5-dichlorophenylmercaptomethyl) dithiophosphate
Phorate	*O,O*-Diethyl *S*-ethylthiomethyl phosphorodithioate
Phosdrin	Dimethyl 1-carbomethoxy-1-propen-2-yl phosphate
Phosphamidon	Dimethyl diethylamido-1-chlorocrotonyl-(2) phosphate
Rhothane	See DDD
Rogor	See dimethoate

(Table Continues)

TABLE 4.1 (*Continued*)

COMMON NAME	CHEMICAL NAME
Ronnel	*O,O*-Dimethyl *O*-2,4,5-trichlorophenyl phosphorothioate
Rotenone	Cube and derris roots
Sevin	1-Napthyl-*N*-methylcarbamate
Simazine	2-Chloro-4,6 bis(ethylamino)-*S*-triazine
Strobane	Mixture of chlorinated terpenes
Sulfotep	Bis-*O,O*-diethylphosphorothionic anhydride
Sumithion	*O,O*-Dimethyl *O*-3-methyl-4-nitrophenyl phosphorothioate
Systox	See demeton
2,4,5-T	2,4,5-Trichlorophenoxy acetic acid
Telodrin	Octachlorohexahydromethanoisobenozofuran
TEPP	Bis-*O,O*-diethylphosphoric anhydride
2,4,5-T ester	Butoxyethanol
Thiodan	6,7,8,9,10,10-Hexachloro-1,5,5*a*,6,9,9*a*-hexahydro-8,9-methano-2,4,3-benzodioxathiepin-3-oxide
2,4,5-TP	2,4,5-Trichlorophenoxy propionic acid
Toxaphene	Essentially a mixture of isomers of octachlorocamphene
Trithion	8-(*p*-Chlorophenylthiomethyl)*O,O*-diethyl phosphorodithioate
Zinophos	*O,O*-Diethyl *O*-(2-pyrazinyl) phosphorothioate

Because of their chemical complexity, it is usually not convenient to identify pesticides by chemical structures; thus many pesticides are referred to by their trade names. Table 4.1 lists the common names and chemical names of pesticides currently in use. This list is included so that the reader can refer to it to determine the chemical structure of a particular compound if needed. It will also allow the reader to determine whether a particular pesticide fits in one of the three general categories outlined previously.

As mentioned earlier, the most thorough environmental pesticide studies have been performed for DDT, which was first synthesized in 1874 and was widely used during World War II for controlling malaria. In the early days of its use the persistence of DDT in the environment was its principle attraction. Figure 4.2 shows the production of DDT since 1950, indicating that the manufacture of DDT reached 188,000,000 lb in 1964, the highest production level of any pesticide in the United States. Figure 4.2 also shows that DDT production declined rapidly until 1972, after which time production was essentially stopped. Figure 4.3 shows the production levels of aldrin and dieldrin during the period from 1965 to 1974 to allow comparison with that of DDT. These pesticides, as will be discussed later, have also been essentially removed from the market because of their potential for

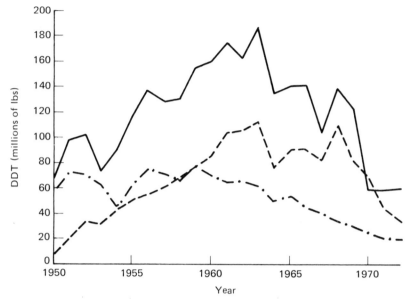

FIGURE 4.2   DOMESTIC PRODUCTION (——), CONSUMPTION (— · —), AND EXPORTS (---) OF DDT FROM 1952 TO 1972.   (From Council of Environmental Quality, 1981.)

inflicting environmental damage.   However, even at the point of maximum production in 1966, only about 20 million lb yr^{-1} of dieldrin was synthesized.   The data for aldrin and dieldrin are combined because aldrin degrades very rapidly to dieldrin in the environment.

The production of other pesticides has been cyclic in nature, depending on the demand for them.   After the pesticides DDT and aldrin were shown to be environmentally harmful, the production of other pesticides was

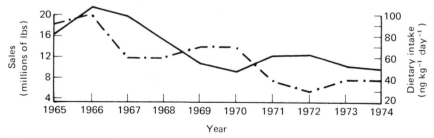

FIGURE 4.3   DOMESTIC SALES OF ALDRIN AND DIELDRIN (——) AND FRACTION OF DIETARY INTAKE (— · —) FROM 1965 TO 1974.   (From Council on Environmental Quality, 1981.)

increased to replace them. The overall trend in pesticide synthesis has been to shorter-lived compounds, which tend to be more toxic. For example, DDT, aldrin, dieldrin, Heptachlor, and Toxephene, which are all chlorinated hydrocarbons, exhibit low toxicities in mammals, are not very target specific, and are long lived in the environment. However, pesticides such as Parathion and Methyl parathion are very easily degraded in the environment and, hence, are quite short lived. These particular pesticides, however, are very toxic to mammals, thus greater care is required in their handling and application.

The ability of pesticides to cause water quality problems is directly related to their longevity in aqueous systems. The more persistent the compound, that is, the longer it resists degradation, the more likely it is to be incorporated into biological systems. Therefore, we should examine the ability of these compounds to be degraded and some of the factors that regulate the degradation process.

B. Chlorinated Hydrocarbons

The first class of pesticides in our categorization is the chlorinated hydrocarbons. These, as the name implies, are hydrocarbons with chlorine substituted at various points on either a chain or ring structure. The amount of chlorine substitution and the exact location of the chlorine molecule is integral to the ability of the compound to resist degradation in the environment. As a general rule, it has been shown that the more chlorine substitution, the better the chemical is as a pesticide and also the more resistant it is to degradation.

The best-documented example of the degradation of a chlorinated hydrocarbon is DDT, or 1,1,1-Trichloro-2,2-bis(p-chlorophenyl)ethane. This compound has already been discussed earlier with respect to its production and use in the environment. Figure 4.4 shows the DDT molecule and two possible breakdown products. It can be seen that microorganisms break down DDT to DDD, whereas mammals and insects break down the molecule to DDE. In the case of the DDD molecule, a single chlorine has been replaced by a hydrogen. In the case of the DDE molecule, a carbon double bond has been formed. It may not seem that these molecules have changed much, but their behavior in the environment is severely affected by the small molecular changes that accompany degradation. Both DDD and DDE are toxic and relatively long lived in the environment, as is DDT. Thus the breakdown products of DDT can pose water quality problems similar to the parent DDT molecule.

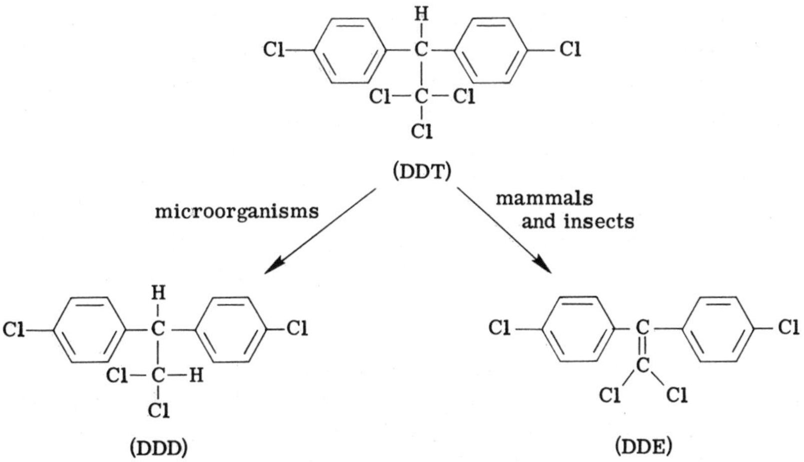

(DDT)

microorganisms

mammals
and insects

(DDD)

(DDE)

FIGURE 4.4   DEGRADATION PATHWAYS OF DDT.

Figure 4.5 shows the breakdown of another chlorinated hydrocarbon, Lindane. The complex ring structure of Lindane can be broken down in either one of two ways, i.e., to 2,4,5-trichlorophenol or 2,3,5-trichlorophenol. These compounds are known to cause taste and odor problems in water. Once again, it can be seen that the difference in the resultant compounds is merely the location of the chlorine on the ring structure. Many of the degradation pathways of chlorinated hydrocarbons have been described. Most of the pesticides in use today are known to degrade, and

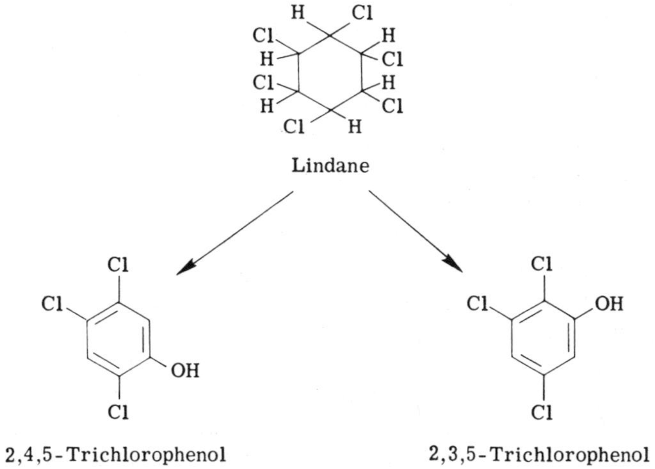

Lindane

2,4,5-Trichlorophenol

2,3,5-Trichlorophenol

FIGURE 4.5   DEGRADATION PATHWAYS OF LINDANE.

their resultant metabolites have been defined. However, from a water quality perspective, the occurrence of metabolites is not as important as the rate at which they are produced. In general, chlorinated hydrocarbons cause severe water quality problems because of very slow degradation rates. Kuhr, Davis, and Taschenberg (1972) have studied vineyard soil that received DDT applications constantly for 24 yr, and they showed that the DDT remained within the top 3 in. of soil and did not move into the ground water. They also showed that approximately 50% of the DDT degraded within 6 yr, 67% degraded after 12 yr, and the only breakdown product found in the soil was DDE.

Mitchell (1974) approximated decomposition times for certain chlorinated hydrocarbons, e.g., DDT and dieldrin, to be approximately 3 yr, whereas Chlordane takes approximately 11 yr. These values are approximations because the type of environment affects the degradation rate. Oloffs, Albright, and Szeto (1972) have suggested certain environmental behavior patterns that regulate the degradation of chlorinated hydrocarbons. They note that the solubility of a compound is a prime regulator of its tendency to be broken down. In particular, they suggest the following general rules.

(1)   If a pesticide is present in concentrations greater than its saturation level, the pesticide will accumulate at the air–water interface and evaporate rapidly to the atmosphere. This, of course, accounts for the global transportation of many pesticides.

(2)   If the pesticide is present in concentrations less than its saturation level, then it remains in the water.

(3)   The presence of other materials can affect the solubility of pesticides in water, keeping them in solution even at high concentrations.

Other work by Oloffs *et al.* (1973) showed that chlorinated pesticides can move directly into the sediments of a water system. Table 4.2 presents a study of Chlordane, Lindane, and DDT in a sediment–water system, showing that all of the compounds analyzed except Lindane entered the sediments within 12 weeks after introduction to the water system. It was also observed that Lindane was the only compound metabolized to any significant extent by the microflora. This indicates that some of the chlorinated hydrocarbons (e.g., Chlordane and DDT) are extremely resistant to bacterial metabolism, whereas others (e.g., Lindane) can be metabolized more rapidly.

Rice and Sikka (1973) showed dieldrin to undergo bioaccumulation by phytoplankton. Their data showed that the phytoplankton concentrated dieldrin by 1000 to 15,000 times the ambient concentration in a water column. They also showed that this was less than the concentration factor for DDT and concluded that the difference in accumulation between the

TABLE 4.2   FATE OF 0.025-PPM CHLORINATED HYDROCARBON
COMPOUNDS IN 150-ML WATER SAMPLES FROM FRASER RIVER AND
GEORGIA STRAIT IN THE PRESENCE OF BOTTOM SEDIMENTS FROM
THE SAME SOURCES[a]

| | | RECOVERY (%) | | | |
| | | FRASER RIVER | | GEORGIA STRAIT | |
COMPOUND	INCUBATION TIME (WEEKS)	WATER	SEDIMENT	WATER	SEDIMENT
α-Chlordane	0	100	0.0	100	0.0
	6	0	90.0	0	69.3
	6	0	87.9	0	81.7
	12	0	79.7	0	78.5
	12	0	86.2	0	69.3
γ-Chlordane	0	100	0.0	100	0.0
	6	0	98.8	0	65.8
	6	0	86.6	0	61.0
	12	0	93.3	0	62.7
	12	0	85.1	0	63.8
Lindane	0	100	0.0	100	0.0
	6	0	6.9	15.1	32.3
	6	0	8.8	19.1	27.7
	12	0	3.2	4.6	10.2
	12	0	2.9	6.2	9.7
DDT	0	100	0.0	100	0.0
	6	0	60.6	0	68.6
	6	0	51.3	0	73.7
	12	0	11.9	0	80.5
	12	0	26.1	0	82.0

[a] Incubated at 13°C. No residues could be detected on the silanized glass wool plugs used to stopper the flasks during incubation. Controls (0–12 weeks) were free from detectable residues. (From Oloffs *et al.*, 1973.)

two compounds was due to the difference in their solubilities. DDT is much less soluble in water than dieldrin and would tend to come out of solution rapidly in the presence of phytoplankton.

C. Organophosphorus Compounds

The second general class of pesticides we shall consider is the organophosphorus compounds. As the name implies, these compounds are organically complexed phosphorus molecules, which generally have the structure shown (1). These compounds are normally easily hydrolized and are

$$R{-}O\diagdown_{P\diagup}^{\diagup O}{-}O{-}\text{(organic radical)}$$
$$R{-}O$$

1

therefore somewhat easily degraded. Some compounds in this group have a general specificity and others are very specific with respect to the target organism. Almost all of the compounds in this group have a very high toxicity in mammals. As an example, one of the more commonly used pesticides in this group, Parathion, has been shown to hydrolize slowly, with about 50% of the compound being degraded in 120 days at a pH of 7. Another common pesticide in this group, DDVP, is 50% degraded in 8 h at a pH of 7.

Hurlbert, Mulla, and Willson (1972) studied the effect of the organophosphorus pesticide Dursban on the ecology of freshwater ponds. The insecticide was used for mosquito control in the ponds, and Hurlbert *et al.* studied the subsequent ecological effects. They noted that Dursban killed most of the herbivorous predators, thus causing a large algal bloom in the pond. It also reduced predator insects to levels substantially below those required for mosquito control. The net result was an increase in the number of insects that the pesticide was intended to kill. These findings are not unique to Dursban, however, and are quite common for any broad-spectrum pesticide. Chambers and Yarbrough (1974) studied the resistance of the mosquito fish *Gambusia* to Parathion and Methyl parathion, showing that the mosquito fish could tolerate 40 times more Methyl parathion than Parathion. They also showed that some resistance to these pesticides could be developed in the fish by low-dose experiments.

D. Common Herbicides and Fungicides

The last group that we shall consider is a conglomerate of several pesticides in common use. The most common members of this group are the herbicides 2,4-dichlorophenoxy acetic acid and 2,4,5-trichlorophenoxy acetic acid. These compounds are abbreviated in the literature as 2,4-D and 2,4,5-T and are principally used as weed-killing agents. Much research has been done on the degradation rates of these compounds. Alexander (1961) studied the degradation rates of 2,4-D and 2,4,5-T, and his observations are

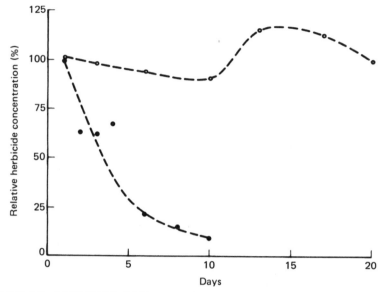

FIGURE 4.6   MICROBIAL DEGRADATION OF 2,4-D (•) AND 2,4,5-T (○) IN SOIL
SUSPENSIONS. (From Alexander, 1961. © 1961 John Wiley & Sons, Inc.)

shown in Fig. 4.6. The addition of extra chlorine on the benzene ring
apparently makes the compound extremely resistant to bacterial degradation.

Generally speaking, the resistance to degradation of pesticides is largely a
function of molecular structure. Some generalizations can be made in that
the most resistant of the pesticides are the chlorinated hydrocarbons, with
the next level of resistance exhibited by the highly substituted phenolben-
zenes. The least resistant, or most easily degraded, pesticides are the
organophosphates. Table 4.3 shows the relative persistence of certain
pesticides in the environment. The chlorinated hydrocarbons are seen to
persist for several years, whereas other pesticides are degraded in months or
weeks.

Table 4.4 shows the EPA limits of organic pesticide concentrations in
public water supplies. Note that some of the compounds, such as the
chlorinated hydrocarbons, are allowed to exist at relatively high levels in
water supplies. This, of course, is not based on the longevity of the
compounds in aqueous systems but on their relatively low toxicity to
mammals. The use of these types of standards for water quality management will be discussed in Chapter 10.

TABLE 4.3 RELATIVE PERSISTENCE AND INITIAL DEGRADATION
REACTIONS OF ELEVEN MAJOR PESTICIDE CLASSES[a]

CHEMICAL CLASS	USE	PERSISTENCE	INITIAL DEGRADATION PROCESS
Chlorinated hydrocarbons	Insecticides	2–5 years	Dehydrohalogenation or epoxidation
Ureas	Herbicides	4–10 months	Dealkylation
Triazines	Herbicides	3–18 months	Dealkylation or dehalogenation
Benzoic acids	Herbicides	3–12 months	Dehalogenation or decarboxylation
Amide	Herbicides	2–10 months	Dealkylation
Phenoxy	Herbicides	1–5 months	Dealkylation, ring hydroxylation, or β-oxidation
Toluidine	Herbicides	6 months	Dealkylation (aerobic) or reduction (anaerobic)
Nitrile	Herbicides	4 months	Reduction
Carbamate	Herbicides, fungicides, insecticides	2–8 weeks	Ester hydrolysis
Aliphatic acids	Herbicides	3–10 weeks	Dehalogenation
Organophosphates	Insecticides	7–84 days	Ester hydrolysis

[a] From Kaufman and Plimmer (1972). © 1972 John Wiley & Sons, Inc.

TABLE 4.4 PERMISSIBLE CONCENTRATIONS
FOR ORGANIC PESTICIDES IN PUBLIC
WATER SUPPLIES[a]

PESTICIDE	PERMISSIBLE CONCENTRATION (μG/L)
Aldrin	17.0
Chlordane	3.0
DDT	42.0
Dieldrin	17.0
Endrin	1.0
Heptachlor	18.0
Heptachlor epoxide	18.0
Lindane	56.0
Methoxychlor	35.0
Organic phosphates and carbamates	100.0
Toxaphene	5.0
2,4-D; 2,4,5-T; 2,4,5-TP	100.0

[a] From Environmental Protection Agency (1980).

IV. PETROLEUM HYDROCARBONS AND
WATER QUALITY

A. Sources of Oil in Marine Environments

Petroleum hydrocarbons are hydrocarbons that are introduced to water systems owing largely to the human activity of reclaiming fossil fuels for energy. The acute need for energy in the world today has resulted in a large output of oil and oil products; thus much of the hydrocarbon material that is withdrawn from the earth in various parts of the world is transported to other parts of the world for refining. A large percentage of this oil, both in the crude form and in various refined forms, is lost to the environment, most of it ending up in the oceans. Table 4.5 shows the estimated sources and magnitudes of petroleum inputs to the marine environment. The table assumes that all of the oil lost at various points in the world eventually ends up in the ocean. Over 6×10^6 ton yr^{-1} of oil are lost, and losses from tankers, oil terminals, transportation, and related sources account for the largest percentage of oil inputs to the oceans. Wilson et al. (1974) estimate that over 600,000 ton yr^{-1} of oil are added to the marine environment by natural seepage. They also showed that most of the seepage occurs in the areas of the world that are the most geologically active, e.g., southern California and Alaska.

Major oil spills pose localized environmental problems and represent a large component of the total oil input to the environment. The rate of oil

TABLE 4.5 NATIONAL ACADEMY OF SCIENCE ESTIMATES
OF INPUTS OF PETROLEUM INTO THE MARINE ENVIRONMENT[a]

SOURCE	INPUT (METRIC TONS YR^{-1})
Tankers, oil terminals, and other transportation— related sources	2,100,000
River and urban runoff	1,900,000
Atmospheric fallout	600,000
Natural seeps	600,000
Industrial wastes	300,000
Municipal wastes	300,000
Coastal refineries	200,000
Offshore production	100,000
Total	6,100,000

[a] From National Academy of Sciences (1974).

loss in the United States is approximately one barrel for every million barrels of oil transported. The United States maintains a fleet of ships that represents approximately 5% of the world's tanker traffic. Most major oil spills have been the result of human error, and one study (Dillingham Corporation, 1970) that considered the characteristics of oil spills larger than 2000 barrels revealed the following information for a 20-year period:

(1)   75% of the oil spills were associated with vessels, mostly tankers;

(2)   90% of the oil spills involved crude oil;

(3)   70% of the oil spills were greater than 5000 barrels (the average spill being 25,000 barrels);

(4)   80% of the oil spills occurred within 10 miles of shore;

(5)   75% of the oil spills were incidents that lasted greater than 5 days, with the median being 17 days;

(6)   80% of the oil spills contaminated less than 20 miles of coastline, the median being 4 miles;

(7)   85% of the oil spills ocurred offshore of recreational areas;

(8)   75% of the oil spills occurred within 25 miles of the nearest port.

These data indicate that most oil spills contain crude oil and occur in areas that are very close to shore. It has been calculated that, with a constant onshore wind, the average oil spill reaches a coastline in slightly over 2 h. This, of course, necessitates quick response times for oil spill containment and cleanup.

The major problem involved in oil spills is the stopping distance of the large tankers. Relatively large tankers, e.g., greater than 400,000 tons, take over 5 miles ($\frac{1}{2}$ h) to stop, assuming the tanker has made an emergency stop and all engines are in full reverse. If the ship were not to reverse the engines, it would take more than an hour to stop. Therefore, the ability of oil tankers to make any sort of emergency maneuver is essentially nonexistent within the span of half an hour.

B. Dispersal and Degradation of Oil in Marine Environments

Assuming, then, that oil is introduced to a natural aqueous environment either by a catastrophic spill, natural seepage, or day-to-day transportation, we must investigate the fate of the oil.

The general processes involved in the dispersal and degradation of oil in aqueous environments are as follows.

(1)   *Spreading* thins the slick out to a few millimeters or less, depending predominantly on the viscosity of the oil.

(2) *Evaporation* is the process by which the low molecular weight compounds with low boiling points are vaporized into the atmosphere. The process is dependent on the viscosity of the oil and the weather conditions at sea.

(3) *Dissolution* is the process by which the low molecular weight compounds travel vertically and are lost to the water column. It is governed by many parameters, including the viscosity of the oil and the weather conditions at sea.

(4) *Emulsification* is the process by which the oil is held in suspension in the water.

(5) *Autooxidation* is a light-catalyzed reaction in which hydrocarbons are oxidized by molecular oxygen to form ketones, aldehydes, and carboxylic acids.

(6) *Microbial degradation,* or the microbial oxidation of oil, is usually performed by bacteria, actinomycetes, fungi, and yeast. This process can proceed either aerobically or anaerobically.

(7) *Sinking* occurs when the density of the oil is increased owing to either evaporation or dissolution. The residual oil then sinks to the bottom and is incorporated in the sediments of the system.

(8) *Resurfacing* occurs when the mass of oil has been reduced to such a degree that it can be refloated from the sediments.

The overall cycle of carbon when introduced to a system as a petroleum hydrocarbon is shown in Fig. 4.7, and it can be seen that part of the carbon is evaporated and joins the atmospheric carbon pool. The residual material, which is extremely resistant to degradation, turns into an emulsion and subsequently into tar. Most of the hydrocarbons in the oil undergo complex autooxidation and biological oxidation processes and remain dispersed or become incorporated in the sediments. They can also be assimilated by marine organisms and undergo biomagnification.

The two most important processes from an environmental point of view, then, are autooxidation and biological degradation. These processes probably account for the major part of the transformation of petroleum hydrocarbons in natural systems. We shall see that most of the data on degradation rates have been collected in marine environments. Extrapolation to freshwater systems, however, should not be difficult owing to the ubiquity of microflora.

1. Autooxidation

Autooxidation deals with autocatalytic reactions that are usually catalyzed by sunlight. The photochemical reactions liberate free radicals that subsequently oxidize other compounds in the petroleum to generate many breakdown products. Dean (1968) suggested that the sulfur and organo-

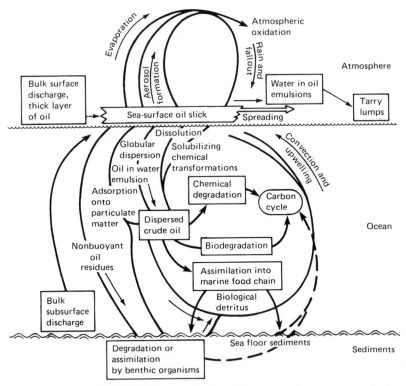

FIGURE 4.7   PROCESSES INVOLVED IN THE FATE OF SPILLED CRUDE OIL IN MARINE ENVIRONMENTS. (From Burwood and Speers, 1974.)

metallic compounds in crude oil play important roles in its oxidation. In particular, the organometallic compounds can act as catalysts for free-radical reactions. The products generally observed in these photo-driven auto-catalytic reactions are acids, carbonyl compounds, alcohols, peroxides, and sulfoxides. Burwood and Speers (1974) suggested a pathway for the auto-oxidation of crude oil, which is shown in Fig. 4.8. This pathway essentially suggests that the process is initiated by photoactivation and results in the formation of free radicals. The free radicals then react with sulfur compounds to form various end products, ultimately resulting in aqueous dissolution.

2. Biological Oxidation

Biological reactions involved in the degradation of petroleum are also important in natural systems. In most instances, biological degradation is

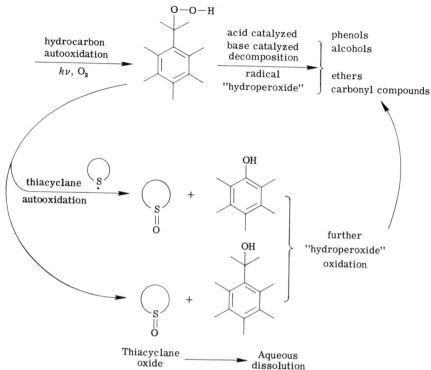

FIGURE 4.8   SUGGESTED ROUTES FOR THE AUTOOXIDATION OF CRUDE-OIL HYDROCARBONS AND THIACYCLANE. (From Burwood and Speers, 1974.)

the predominant mechanism involved. In general, it can be assumed that when crude oil is discharged into an aqueous system, those fractions with boiling points less than 370°C will evaporate from the system in a matter of days. This leaves biological and autocatalytic decomposition to operate on the remaining fraction. An important concept to remember at this point is that biological decay usually involves only specific compounds. Crude oil is a complex mixture of hydrocarbons, many of which are toxic to microorganisms and many of which are easily tolerated and degraded. Thus, we shall see that as a microbial population attacks crude oil, certain fractions will be utilized preferentially and certain fractions will remain undisturbed.

Mitchell (1974) listed the general hierarchy of hydrocarbons with respect to preference for microbial degradation. He suggests that alkanes are more readily degraded than aromatic hydrocarbons and that within the alkanes straight-chain compounds are more susceptible to microbial action than compounds with branching chains. Compounds with chain lengths of 10 to

18 carbon atoms are the most rapidly oxidized. Methane, ethane, and propane are attacked by only a few highly specialized organisms, and the more refractory materials, e.g., waxes or compounds containing more than 30 carbon atoms, are very insoluble and therefore very resistant to degradation.

Walker and Colwell (1974) investigated the degradation of a "model" petroleum as a function of various temperatures. They synthesized the model petroleum by adding known amounts of specific hydrocarbons and then introduced an inoculum from the Chesapeake Bay area. Table 4.6 shows the percentage of various hydrocarbons remaining in the model petroleum after microbial degradation. These particular data were taken at 5°C, but rates at colder temperatures were not appreciably different. They concluded that bacterial oxidation is quite important at temperatures below 5°C, and the group of microorganisms operative in their studies included *Vibrio, Aeromonas, Pseudomonas,* and *Acinetobacter.* They also showed that only certain compounds in the hydrocarbon mixture were collectively removed by the microflora during the process of degradation.

Atlas and Bartha (1973) isolated oil-degrading microorganisms from Raritan Bay in New Jersey. They found that the number of organisms varied between 20 and 3400 l^{-1} depending on the level of oil pollution in the area. The occurrence of oil-degrading microorganisms in a particular situation can apparently be increased depending on the amount of residual oil present. Reisfeld, Rosenberg, and Gutnick (1972) isolated bacteria capable of emulsifying oil, and they found that bacteria of the genus *Arthrobacter* are the most efficient in this process. They were able to develop the bacteria such that 65% of the test oil was converted into non-benzene-extractable forms. They were also able to disperse oil in less than 1 day after inoculation with *Arthrobacter.* Once again, it should be emphasized that microorganisms may be capable of emulsifying an oil, that is, putting it into suspension in the water column, and not degrading the majority of the components in the oil. When dealing with the degradation of oil one has to be very careful to define whether the oil is in fact being degraded or simply converted into an emulsified form that is suspended in the water column.

It can be seen from the data presented that actual degradation rates of crude oil are difficult to obtain. Certain components of crude oil have been known to be removed by certain microflora at a certain rate, and some data are available for emulsification rates. None of these values are especially applicable, however, when dealing with the actual loss of hydrocarbons in an aqueous system, because the ability of the natural microflora to break down crude oil is often dependent on factors other than the oil itself. For instance, a large oxygen supply is required for degradation to occur, and the available

TABLE 4.6 BACTERIAL UTILIZATION OF MODEL PETROLEUM AT 5°C[a]

| | | HYDROCARBONS REMAINING AFTER MICROBIAL DEGRADATION (%) | | | | | | | | |
| | | COLGATE CREEK SEDIMENT | | COLGATE CREEK WATER | | EASTERN BAY SEDIMENT | | EASTERN BAY WATER | |
HYDROCARBON	CONTROL	SALTS	WATER	SALTS	WATER	SALTS	WATER	SALTS	WATER
Cyclohexane	100.00	5.42	7.27	3.77	14.30	2.72	5.61	6.63	59.99
n-Decane	100.00	51.44	61.31	41.55	57.93	35.65	65.16	50.78	63.76
Cumene	100.00	2.99	5.61	2.77	11.11	1.96	3.03	4.76	56.12
n-Undecane	100.00	59.86	62.18	47.59	55.29	34.87	59.80	52.00	67.12
n-Dodecane	100.00	61.62	62.64	52.81	63.51	34.96	58.09	50.67	64.69
n-Tridecane	100.00	61.64	62.49	47.52	55.10	34.46	56.49	49.15	62.26
Naphthalene	100.00	2.40	4.08	1.77	9.95	1.34	2.35	3.42	55.17
n-Tetradecane	100.00	61.48	62.23	47.31	55.06	34.27	55.71	48.67	61.28
n-Pentadecane	100.00	61.41	61.99	47.27	55.00	34.26	55.31	48.28	60.31
n-Hexadecane	100.00	61.93	62.06	47.59	55.35	34.61	55.47	48.86	60.72
Pristane	100.00	61.66	62.39	47.68	54.77	34.65	55.46	45.61	54.77
n-Heptadecane	100.00	61.91	61.38	47.33	55.30	34.51	54.96	48.98	61.00
n-Octadecane	100.00	60.00	61.88	48.52	56.37	35.62	55.95	51.01	62.32
n-Nonadecane	100.00	61.37	63.59	48.57	56.26	35.80	55.66	50.62	61.79
n-Eicosane	100.00	61.94	64.60	50.13	56.98	36.45	56.33	51.36	62.36
Phenanthrene	100.00	55.40	60.99	40.73	52.86	31.76	53.07	38.80	53.09
1,2-Benzanthracene	100.00	38.73	59.23	42.35	52.05	33.27	52.35	34.30	46.77
Perylene	100.00	52.32	60.70	40.82	51.54	31.81	52.13	31.85	44.97
Pyrene	100.00	56.17	59.76	42.96	52.28	32.20	51.75	34.01	46.21

[a] From Walker and Colwell (1974).

TABLE 4.7   CHEMICAL COMPOSITION OF SWAN HILLS OIL AFTER 308 DAYS IN CONTACT WITH SOIL PLUS VARIOUS AMENDMENTS[a]

	COMPOSITION OF OIL (%)[b]				
CRUDE OIL FRACTION	BARREL	SOIL	SOIL AND BACTERIA	SOIL AND FERTILIZER	SOIL, BACTERIA, AND FERTILIZER
Asphaltenes, soluble	3.00	1.35	1.52	2.01	2.58
Asphaltenes, insoluble	0.24	12.71	12.42	20.69	21.31
Saturates	62.10	54.74	56.36	39.30[c]	37.91[c]
Aromatics	23.20	17.94	18.11	20.90	21.14
NSO[d] soluble	5.60	6.80	7.63	9.06	9.98
NSO insoluble	5.70	6.42	4.77	8.04	7.27

[a] From Jobson et al. (1974).
[b] Average of values from four replicate plots.
[c] Significantly different from the unamended oil treatment values at both the 95 and 99% confidence levels.
[d] NSO: organic compounds containing nitrogen, sulfur, and oxygen.

nitrogen and phosphorus is generally insufficient in open ocean waters to maintain a population of microflora capable of degrading large oil spills. The oil, of course, represents a large carbon supply, but oil contains little nitrogen and phosphorus. Many studies have shown that the addition of nitrogen and phosphorus to areas containing high concentrations of oil facilitates the degradation process. Jobson et al. (1974) studied the degradation of oil as a function of bacteria and nutrients in soil systems and showed that the addition of nitrogen and phosphorus as well as bacteria to oil-soaked sediments stimulates the flora that are capable of degradating the oil. They also demonstrated that alkanes are the principal components of oil that are easily removed by bacteria. Table 4.7 shows the results of their studies, and the need for adding nutrients such as nitrogen and phosphorus is readily apparent. This has been performed in other areas where severe oil spills have occurred, and increased rates of degradation have been observed.

V.   DETERGENTS

Detergents can be considered to be artificial soaps, for the largest use and subsequent input to water systems comes from their use as cleaning agents in domestic households. Although the chemical structure of detergents and their resistance to degradation is the main theme of this chapter, it is interesting to reflect on the history of detergent use in the United States. This history makes a very nice case study of how water quality problems can evolve and the subsequent inability of communities to cope with them.

For hundreds of years soaps have been used throughout the world to handle any cleaning chore. It was well known that compounds called "soap" were capable of cleaning clothes and other surfaces that had been soiled. The general soap compounds were synthesized from fat, and the typical structure is shown (2). This long-chain compound is easily degraded

$$H-\underset{\underset{H}{|}}{\overset{\overset{H}{|}}{C}}-\underset{\underset{H}{|}}{\overset{\overset{H}{|}}{C}}-\underset{\underset{H}{|}}{\overset{\overset{H}{|}}{C}}-\underset{\underset{H}{|}}{\overset{\overset{H}{|}}{C}}-\underset{\underset{H}{|}}{\overset{\overset{H}{|}}{C}}-\underset{\underset{H}{|}}{\overset{\overset{H}{|}}{C}}-\underset{\underset{H}{|}}{\overset{\overset{H}{|}}{C}}-\underset{\underset{H}{|}}{\overset{\overset{H}{|}}{C}}-\underset{\underset{H}{|}}{\overset{\overset{H}{|}}{C}}-\underset{\underset{H}{|}}{\overset{\overset{H}{|}}{C}}-\underset{\underset{H}{|}}{\overset{\overset{H}{|}}{C}}-C\overset{\nearrow O}{\searrow_{Na}}$$

2

by bacteria and has an environmental residence time of less than a day. The occurrence of soap in water systems, therefore, seldom posed any problem.

During World War I fats were difficult to obtain in Germany. Scientists in that country started searching for synthetic cleaners, and Fritz Gunther developed the first artificial soap in 1916. It was a diisopropylnapthalene-sulfonate compound that worked very well in place of natural soaps.

In 1932 the first household detergent marketed in the United States was developed in response to a need for a soap substitute in those areas of the country with hard water. It was well known that soaps, being purely an organic compound, would chelate the minerals in the water to form a mineral–organic curd. This curd would precipitate on sink bowls as a grey scum and would also cause fabrics to turn grey. The news that Germany had found an artificial detergent spurred the U. S. soap industry to develop its own. The first detergent marketed in the United States was called Dreft, a product of Proctor and Gamble.

In general, artificial detergents contain three basic compounds: a surfactant or wetting agent, a builder or complexing agent for the minerals in water, and an alkali. The builder is generally a phosphate, and in Chapter 5 we shall explain how phosphate from detergents can accelerate the eutrophication process in natural waters. In this chapter, however, it is the surfactant, or wetting agent, that is the water quality problem of interest.

Originally the compounds used as surfactants in detergents were a very diverse group; they did, however, tend to have the same general configuration. The compounds all had both hydrophylic and hydrophobic polarity, i.e., one end of the molecule could be dissolved in the water and the other could not. Foreign material is adsorbed on one end of the molecule, whereas the other end remains insoluble. The foreign material can then be removed during the washing cycle. The basic properties of detergents are

(1)   concentration at surfaces,
(2)   lowering of surface tension in the solution,
(3)   formation of aggregate ions in solution.

The compounds that are used as detergents can be split into three basic groups:

(1)   nonionic, no charge associated with the molecule;
(2)   anionic, negative charge associated with the molecule;
(3)   cationic, positive charge associated with the molecule.

The cationic detergents have found only specialized use in the United States. Owing to their low frequency of application, cationic detergents are not considered to be a water quality problem. The nonionic detergents are widely used, but there is a lack of data on their behavior in aqueous systems. However, the amount of information available suggests that they behave in a manner similar to the anionic compounds. We shall therefore use the anionic detergents as a model in this chapter, for most of the available data applies to anionic detergents. Their use is widespread in the United States, and they represent the largest contribution to the environmental pool of organic surfactants.

Most anionic detergents are alkylarylsulfonates, and the most commonly used of this type was alkylbenzenesulfonate (ABS). This compound is very resistant to microbial degradation because of its branched-chain structure. The structural formula (3) shows an example of an ABS molecule, and the

$$NaSO_3 - \bigcirc - \overset{\overset{H}{|}}{\underset{\underset{H}{|}}{C}} - \overset{\overset{H}{|}}{\underset{\underset{H}{|}}{C}} - \overset{\overset{H}{|}}{\underset{\underset{H}{|}}{C}} - \overset{\overset{H}{|}}{\underset{\underset{H}{|}}{C}} - \overset{\overset{H}{|}}{\underset{\underset{H}{|}}{C}} - \overset{\overset{H}{|}}{\underset{\underset{H}{|}}{C}} - \overset{\overset{H}{|}}{\underset{\underset{H}{|}}{C}} - \overset{\overset{H}{|}}{\underset{\underset{H}{|}}{C}} - \overset{\overset{CH_3}{|}}{\underset{\underset{CH_3}{|}}{C}} - \overset{\overset{H}{|}}{\underset{\underset{H}{|}}{C}} - CH_3$$

3

persistence of this compound in water systems caused aesthetic as well as public health problems. Because of its persistence and surface-tension-lowering capacity, many foaming problems developed when ABS reached high concentrations in water systems. Treatment plants employing aeration generated tremendous foaming problems, and in some severe cases tap water would foam in the sink owing to the presence of ABS. The literature abounds with references describing the toxicity of ABS and its resistance to degradation; most of this work was carried out in the 1950s and early 1960s. Once informed of the environmental problems involved with ABS, the soap and detergent industry substituted a similar compound that did not have a branched chain.   The compound substituted was a linear alkylsulfonate (LAS), and the structure of the molecule is shown (**4**).   This molecule

4

is easily degraded by bacteria and hence does not persist in the environment.   It is also just as effective a surfactant for use in detergents.   Therefore, after 1965 very little ABS was used in detergents.   It should be noted, however, that, although LAS is not as persistent in the environment as ABS, it is much more toxic to aquatic organisms.

Commercial detergents may contain any one of approximately five forms, or isomers, of LAS.   It is therefore difficult to classify these compounds with respect to toxicity, not only because there are so many forms, but because their breakdown products are also variable.   All LAS surfactants, however, have been shown to be toxic to aquatic life at very low concentrations.   Swedmark *et al.* (1971) found certain fin fish, bivalves, and decapods to be very susceptible to low concentrations of LAS.   They found both anionic and nonionic detergents to be very toxic and determined that the most active species, i.e., fish, tend to be the most susceptible.   Surfactants probably affect the ionic balance in fish and osmoregulation in bivalves.

Hirsch (1963) showed that the toxicity of LAS compounds is a function of chain length.   Figure 4.9 shows that, as the chain length becomes longer, the compound exhibits a greater toxicity to fish, that is, a lower dose is required for death.   It has also been shown that the toxicity of these compounds is a function of both the oxygen concentration and the hardness of water systems.   An unfortunate aspect of this problem is that there is no trend to these affects, and consequently some compounds are more toxic in soft water, whereas some are more toxic in hard water.   The dependency on oxygen concentration also exhibits the same randomness.   Abel (1974) summarized the toxicity of several surfactants to various fish.   Table 4.8 shows the results of these studies.   It appears that once again the values are variable, and they probably reflect the different aqueous systems in which

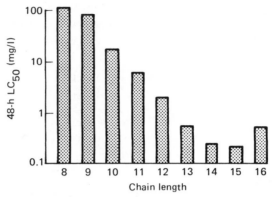

FIGURE 4.9   THE INFLUENCE OF LAS CHAIN LENGTH ON DETERGENT TOXIC-
ITY.   The 48-h LC_{50} of ABS to *Idus idus* (L.) is plotted on a logarithmic scale against
the number of carbon atoms in the hydrocarbon chain.   (From Hirsch, 1963.)

TABLE 4.8   ACUTE TOXICITY OF SURFACTANTS TO TEST
SPECIES OF FISH UNDER VARYING CONDITIONS[a]

SURFACTANT	TEST SPECIES	RESULT
ABS	*Salmo gairdneri* (Richardson)	10-day LC_{50} of 5 mg/l
ABS	*Pimephales promelas* (Rafinesque)	96-h LC_{50} between 3 and 9 mg/l
ABS	*Lepomis macrochirus* (Rafinesque)	96-h LC_{50} of 17 mg/l
ABS	*Lepomis gibbosus* (L.)	96-h LC_{50} of 22 mg/l
ABS	*Fundulus heteroclitus* (L.)	96-h LC_{50} of 6.8 mg/l
ABS	*Mugil cephalus* (L.)	96-h LC_{50} of 3.1 mg/l
ABS	*Pseudopleuronectes americanus* (Walbaum)	96-h LC_{50} of 2.5 mg/l
ABS	*Anguilla rostrata* (le Sueur)	96-h LC_{50} of 2.3 mg/l
ABS	*Menidia menidia* (L.)	96-h LC_{50} of 2.1 mg/l
ABS	*Notropis atherinoides* (Rafinesque)	96-h LC_{50} of 7.4 mg/l
ABS	*Pimephales notatus* (Rafinesque)	96-h LC_{50} of 7.7 mg/l
ABS	*Lepomis macrochirus* (Rafinesque)	96-h LC_{50} of 8.2 mg/l
ABS	*Campostoma anomalum* (Rafinesque)	96-h LC_{50} of 8.9 mg/l
ABS	*Notropis stramineus* (Cope)	96-h LC_{50} of 9.0 mg/l
ABS	*Erycymba buccata* (Cope)	96-h LC_{50} of 9.2 mg/l
ABS	*Notropis ardens* (Cope)	96-h LC_{50} of 9.5 mg/l

TABLE 4.8   *(Continued)*

SURFACTANT	TEST SPECIES	RESULT
ABS	*Pimephales promelas* (Rafinesque)	96-h LC_{50} of 11.3 mg/l
ABS	*Notropis cornutus* (Mitchill)	96-h LC_{50} of 17.0 mg/l
ABS	*Cyprinis carpio* (L.)	96-h LC_{50} of 18.0 mg/l
ABS	*Ictalurus melas* (Rafinesque)	96-h LC_{50} of 22.0 mg/l
ABS	*Gadus morrhua* (L.)	96-h LC_{50} of 3.5 mg/l
ABS	*Pleuronectes flesus* (L.)	96-h LC_{50} of 6.5 mg/l
LAS	*Lepomis macrochirus* (Rafinesque)	96-h LC_{50} of 3 mg/l
LAS	*Lepomis macrochirus* (Rafinesque)	96-h LC_{50} of 0.6 mg/l
LAS	*Notropis atherinoides* (Rafinesque)	96-h LC_{50} of 3.3 mg/l
LAS	*Lepomis macrochirus* (Rafinesque)	96-h LC_{50} of 4.0 mg/l
LAS	*Pimphales promelas* (Rafinesque)	96-h LC_{50} of 4.2 mg/l
LAS	*Notropis cornutus* (Mitchill)	96-h LC_{50} of 4.9 mg/l
LAS	*Ictalurus melas* (Rafinesque)	96-h LC_{50} of 6.4 mg/l
LAS	*Lepomis macrochirus* (Rafinesque)	Lethal threshold conc. of 1.6–3.1 mg/l
LAS	*Gadus morrhua* (L.)	96-h LC_{50} of 1 mg/l
LAS	*Pleuronectes flesus* (L.)	96-h LC_{50} of 1.5 mg/l
Alkyl sulphate	*Pimephales promelas* (Rafinesque)	96-h LC_{50} of 5–6 mg/l
Polyoxyethylene ether	*Salmo salar* (L.)	Lethal threshold conc. of 2.5 mg/l
Polyoxyethylene ether	*Salmo salar* (L.)	Lethal threshold conc. of 37 mg/l
Polyoxyethylene ester	*Lepomis macrochirus* (Rafinesque)	96-h LC_{50} of 37 mg/l
Polyoxyethylene ester	*Salmo salar* (L.)	Lethal threshold conc. of 22 mg/l

[a] From Abel (1974).

the tests were run.   The data show that the Environmental Protection Agency's drinking water standard for ABS compounds of 0.5 mg l^{-1} seems to be a good standard.   However, the World Health Organization's standard of 1 mg l^{-1} for water not used for drinking seems a little high.   This standard is likely to allow ABS concentrations to exceed the critical levels for the juvenile stages of many species of fish.

VI. POLYCHLORINATED BIPHENOLS

Polychlorinated biphenols (PCBs) have recently become a grave environmental concern. These organics are very similar to the pesticides discussed earlier, that is, they are aromatic compounds consisting of 12 to 68% substituted chlorine. These compounds have an exceptionally high stability and a very low flammability. They have been used industrially in large amounts since 1920 and have found use in the manufacturing of plastics, wrapping paper, pesticides, ink, paint, tires, and other products. PCBs are extremely good fire retardants, so they have also been used in the fabrication of electrical equipment. The use of PCBs is presently restricted to electrical equipment, owing to recent concerns over the persistence of the compound in the environment.

PCBs bioaccumulate in a manner similar to DDT, and one of the reasons PCBs were not identified sooner as an environmental problem is their structural similarity to DDT and its related breakdown products. Morgan (1972) found 0.02 ppm PCB to be toxic to *Daphnia,* whereas a 2.0 ppm concentration was toxic to guppies. Greichus, Greichus, and Emerick (1973) found that freshwater fish can contain PCB at 12 times the sediment concentrations and also that PCB levels are higher than many insecticides studied in both sediments and fish. Fisher, Graham, and Carpenter (1973) showed that phytoplankton isolated from offshore and coastal waters have variable resistance to PCB concentrations. That is, offshore species are very sensitive to PCB, whereas coastal species are not as sensitive. They also noted that coastal and offshore species are in fact both sensitive to high PCB levels, but adaptation in the coastal area owing to higher ambient levels of PCB allowed the coastal species to develop an apparent resistance.

As with all of the organics discussed in this chapter, PCBs are extremely resistant to bacterial degradation, and the ability of the compound to be degraded is a function of chlorine substitution. Ahmed and Focht (1973) showed that *Achromobacter* can degrade PCB quite readily, and Fig. 4.10 shows examples of oxygen uptake rates of test bacteria metabolizing various PCBs. It should be noted that the more substitution of chlorine on the phenol rings, the more resistant the compound is to degradation. This is shown in the figure by lower uptake rates for the dichlorobiphenyls than for the chlorobiphenyls.

PCBs have been isolated from many water systems around the world, including open ocean water, freshwater lakes, and stream systems. Although the sources of PCBs are almost always industrial, the transportation mode of these compounds is not clear. Martell, Rickert, and Siegel (1975) found PCBs in residential lakes that are only 10 years old and that have no

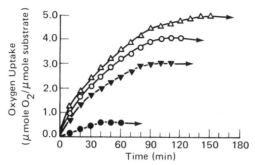

FIGURE 4.10   RATES OF OXYGEN UPTAKE BY WASHED CELL SUSPENSIONS OF PCB-GROWN *ACHROMOBACTER* COMETABOLIZING VARIOUS SUBSTRATES. All rates are corrected for endogenous respiration. △, *o*-chlorobiphenyl; ○, *m*-chlorobiphenyl; ▼, 4,4′-dichlorobiphenyl; ●, 2,2′-dichlorobiphenyl. (From Ahmed and Focht, 1973.)

sewage or industrial waste inputs. They postulated that the PCB concentrations were caused by runoff water from a nearby urban area. Runoff water from streets and parking lots that contain bits of rubber compounds or plasticizers is apparently sufficient to build up PCB concentrations in freshwater systems.

PCBs are now proving to be a more severe environmental problem than

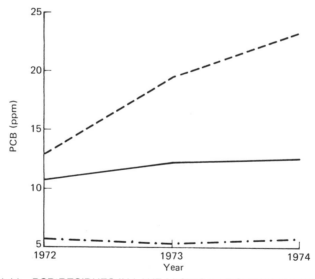

FIGURE 4.11   PCB RESIDUES IN LAKE MICHIGAN FISH FROM 1972 TO 1974: ---, lake trout; ——, Coho salmon; — · —, chubs. (From Council of Environmental Quality, 1981.)

initially supposed. When compared with similar compounds such as DDT, the residence time of PCB is much longer. The estimated inputs of PCBs are currently many times those of DDT. However, as previously mentioned, the production of PCB was curtailed in 1971, so that its use in products is currently very limited. The problem, however, is that the monitoring of many environmental parameters has shown PCB levels to be remaining high. Figure 4.11, for instance, shows PCB residues in Lake Michigan fish, and it can be seen that the levels have risen or remained constant since the early 1970s. If this record of environmental persistence is compared with that of DDT shown earlier, it can be seen that DDT has apparently been removed from environmental systems since its use was curtailed, whereas PCB has not. Data such as these have prompted the Federal Food and Drug Administration to issue an interim safety standard of 5.0 ppm PCB in any food. The Environmental Protection Agency has issued limits on PCB discharges, deciding that environmental levels should remain below 0.01 ppm at all times.

VII.   HUMIC SUBSTANCES

The refractory organics to be discussed in this section are somewhat different than those described previously. This is because by themselves they do not represent a toxic component of the environment. We are referring to those compounds classified as humic substances, or naturally occurring organics, which reach water systems by leaching from soil and decaying plant material. The compounds themselves are not toxic and do not pose a threat to water quality in most situations. However, their interaction with certain chemical schemes for water and wastewater treatment can pose environmental problems.

The term "humic" substances includes a large and variable group of organic compounds that are not easily quantified. The total concentration of humic substances in water systems varies from about 1 to 5 g m^{-3} in fresh water to 0.003 to 2.5 g m^{-3} in open ocean water. The exact composition of humic substances, however, is difficult to determine and has usually been treated by reference to certain extraction procedures. That is, the given set of procedures for extracting the organic material from either soil or water dictates the classification of the compounds that are found. For instance, Fig. 4.12 shows the extraction procedure commonly applied to humic substances in soil. The organic mixture in the soil is extracted with sodium hydroxide, after which there are two components: a soluble component and an insoluble residue (which is called humin). The soluble material is then

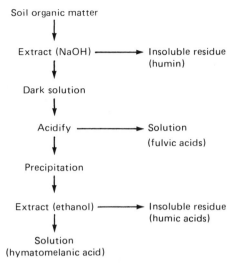

Soil organic matter

Extract (NaOH) ──────► Insoluble residue
                              (humin)

Dark solution

Acidify ──────► Solution
                       (fulvic acids)

Precipitation

Extract (ethanol) ──────► Insoluble residue
                                    (humic acids)

Solution
(hymatomelanic acid)

FIGURE 4.12   PROCEDURE FOR EXTRACTING HUMIC SUBSTANCES FROM
SOIL.

acidified, and the solution phase after acidification is termed fulvic acid.
The precipitated fraction is then extracted with ethanol, and the insoluble
residue from that extraction is called humic acid.   The soluble portion is
classified as hymatomelanic acid.   It should be reemphasized that these
humic compounds, that is, fulvic acid, humic acid, and hymatomelanic
acid, are really only defined by the procedures used to extract them from the
soil complex.   It has been this inability of chemists to actually classify humic
compounds that has caused the problem in defining the reactions involved
when these organic compounds interact with treatment chemicals.
    Schnitzer (1972) showed that there is a difference, albeit a slight one,

TABLE 4.9   ELEMENTAL ANALYSIS OF
HUMIC SUBSTANCES[a]

| | DRY AND ASH-FREE WEIGHT (%) | |
ELEMENT	HUMIC ACIDS	FULVIC ACIDS
C	50-60	40-50
H	4-6	4-6
N	2-6	<1-3
S	0-2	0-2
O	30-35	44-50

[a] From Schnitzer (1972).

between humic acids and fulvic acids in the percentages of the various elements present. Table 4.9 shows an elemental analysis of humic substances. Note that humic acids tend to have a higher percentage of carbon and nitrogen, whereas fulvic acids tend to have a higher oxygen content.

Prakash (1971) showed that the molecular weight of compounds isolated from humic substances varies depending on the location of the source. Table 4.10 shows the variation in molecular weight for four different humic substances, e.g., a soil extract, a river water extract, an extract from a mangrove swamp, and an extract from a coastal embankment. The distribution of molecular weights is quite variable, and even though the humic substances involved reflect plant decay residuals, they apparently vary from system to system. The majority of the compounds found in humic substances from all four situations have molecular weights less than 700.

Now that the composition of humic substances is defined, at least in general terms, we shall see why a water quality problem develops with respect to these compounds. When humic substances reach a water treatment plant, they react with certain treatment chemicals, in particular the oxidants. The most common oxidant used in water treatment is chlorine, which reacts readily with dissolved organics. The reaction is variable, but in general the chlorine concentration is not high enough to completely oxidize the organic material present, and substitution reactions take place. Chlorine atoms are added to certain structures of the humic acid molecules, forming chlorinated organics such as chloroform ($CHCl_3$). The carcinogenicity of chloroform is well known, hence its formation during water treatment processes presents a health problem. A second poblem is created when bromide ion is present, which is common in many waste treatment streams. Chlorine has an oxidation potential sufficient to oxidize bromide ion to bromine, and bromine also reacts with organics in the same manner

TABLE 4.10 MOLECULAR WEIGHT DISTRIBUTION OF HUMIC SUBSTANCES EXTRACTED FROM VARIOUS SOURCES DETERMINED BY THE SEPHADEX GEL-FILTRATION METHOD[a]

MOLECULAR WEIGHT RANGE	SOIL EXTRACT (AQUEOUS) (%)	RIVER WATER (%)	MANGROVE SWAMP (%)	COASTAL EMBAYMENT (%)
<700	43.8	63.0	86.4	68.3
700–1500	56.2	26.8	13.6	23.9
1500–5000	0.0	11.2	0.0	7.8
5000–10000	0.0	0.0	0.0	0.0

[a] From Prakash (1971).

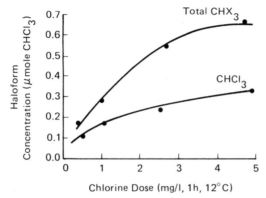

FIGURE 4.13   FORMATION OF HALOFORMS IN WATER OF LOW ORGANIC MATTER CONTENT: HALOFORM CONCENTRATION VERSUS CHLORINE DOSE. [Reprinted from *J. Am. Water Works Assoc.* Vol. 68, No. 8 (March 1976), by permission. © 1976 The American Water Works Association.]

as chlorine.   Hence, a wide range of halogenated organics can be formed by reactions between chlorine and humic substances.

Rook (1976) reported on the formation of haloforms in a water treatment plant in the Netherlands.   He showed that 50 mg l^{-1} of chloroform and over 20 mg l^{-1} of various brominated compounds formed in the Rotterdam treatment plant owing to chlorination.   Figure 4.13 shows the relationship between the amount of organic material present and the chlorine dose for haloform formation.   As more chlorine is added to the water, an increase in halogenated organics is observed.   Therefore, although these refractory, naturally occurring organics do not present a water quality problem in themselves, they can generate a severe water quality hazard upon reaction with water treatment chemicals.

REFERENCES

Abel, P. D. (1974).   Toxicity of synthetic detergents to fish and aquatic invertebrates.   *J. Fish Biol.* **6,** 279.

Ahmed, M., and Focht, D. D. (1973).   Degradation of polychlorinated biphenyls by two species of *Achromobacter. Can. J. Microbiol.* **19,** 47.

Alexander, M. (1961).   "Introduction to Soil Microbiology."   Wiley, New York.

Atlas, R. M., and Bartha, R. (1973).   Abundance, distribution and oil biodegradation potential of microorganisms in Raritan Bay.   *Environ. Pollut.* **4,** 291.

Burwood, R., and Speers, G. C. (1974).   Photo-oxidation as a factor in the environmental dispersal of crude oil.   *Estuarine Coastal Mar. Sci.* **2,** 117.

Chambers, J. E., and Yarbrough, J. D. (1974). Parathion and Methyl parathion toxicity to insecticide resistant and susceptible mosquitofish. *Bull. Env. Contam. Toxicol.* **11**(4), 315.

Council on Environmental Quality (1981). "Environmental Quality." 12th Annual Report of the Council of Environmental Quality. U. S. Government Printing Office, Washington, D. C.

Dean, R. A. (1968). The chemistry of crude oils in relation to their spillage on the sea. *In* "The Biological Effects of Oil Pollution in Littoral Communities," p. 1. Field Studies Council, London.

Dillingham Corporation (1970). Systems Study of Oil Spill Clean up Procedures. Dillingham Corporation, La Jolla, California.

Environmental Protection Agency (1980). Federal Primary Drinking Water Regulations.

Fisher, N. S., Graham, L. B., Carpenter, E. J. (1973). Geographic differences in phytoplankton sensitivity to PCBs. *Nature* **241**(5391), 548.

Greichus, Y. A., Griechus, A., and Emerick, R. J. (1973). Insecticides, polychlorinated biphenyls and mercury in wild cormorants, pelicans, their eggs, food and environment. *Bull. Environ. Contam. Toxicol.* **9**(6), 321.

Hirsch, E. (1963). Strukturelemente von Alkylbenzosulfonaten und ihr Einflub auf das Verhalten von Vischen. *Vom Wasser* **30**, 249.

Hurlbert, S. H., Mulla, M. S., and Willson, H. R. (1972). Effects of an organophosphorus insecticide on the phytoplankton zooplankton, and insect populations of fresh water ponds. *Ecol. Monogr.* **42**, 269.

Jobson, A., McLaughlin, M., Cook, F. D., and Westlake, D. W. S. (1974). Effect of amendments on the microbial utilization of oil applied to soil. *Appl. Microbiol.* **27**(1), 166.

Kaufman, D. D., and Plimmer, J. R. (1972). Approaches to the synthesis of soft pesticides. *In* "Water Pollution Microbiology" (Ralph Mitchell, ed.), pp. 173–206. Wiley (Interscience), New York.

Kuhr, R. J., Davis, A. C., and Taschenberg, E. F. (1972). DDT residues in a vineyard soil after 24 years of exposure. *Bull. Environ. Contam. Toxicol.* **8**, 329.

Martell, J. M., Rickert, D. A., and Siegel, F. R. (1975). PCBs in suburban watershed, Reston, Virginia. *Environ. Sci. Technol.* **9**(9), 872.

Mitchell, R. (1974). "Introduction to Environmental Microbiology." Prentice-Hall, Englewood Cliffs, New Jersey.

Morgan, J. R. (1972). Effects of Aroclor 1242 and DDT on cultures of an alga, daphnid, ostracod, and guppy. *Bull. Environ. Contam. Toxicol.* **8**(3), 129.

National Academy of Sciences (1974). Inputs, fates, and effects of petroleum in the marine environment.

Oloffs, P. C., Albright, L. J., and Szeto, S. Y. (1972). Fate and behavior of five chlorinated hydrocarbons in three natural waters. *Can. J. Microbiol.* **18**, 1393.

Oloffs, P. C., Albright, L. J., Szeto, S. Y., and Lau, J. (1973). Factors affecting the behavior of five chlorinated hydrocarbons in two natural waters and their sediments. *J. Fish. Res. Board Can.* **30**(11), 1619.

Prakash, A. (1971). Terrigenous organic matter and coastal phytoplankton fertility. *In* "Fertility of the Sea" (J. D. Costlow, ed.). Gordon & Breach, New York.

Reisfeld, A., Rosenberg, E., and Gutnick, D. (1972). Microbiol. degradation of crude oil: factors affecting the dispersion in sea water by mixed and pure cultures. *Appl. Microbiol.* **24**(3), 363.

Rice, C. P., and Sikka, H. C. (1973). Fate of dieldrin in selected species of marine algae. *Bull. Env. Contam. Toxicol.* **9**(2), 116.

Rook, J. J. (1976). Haloforms in drinking water. *J. Am. Water Works Assoc.* **68**, 168.

Schnitzer, M. (1972).   Chemical, spectroscopic, and thermal methods for the classification and characterization of humic substances.   Proceedings of the International Meeting on Humic Substances, Nieuwersluis, The Netherlands.

Swedmark, M., Braaten, B., Emanuelsson, E., and Granmo, Å. (1971).   Biological effects of surface active agents on marine animals.   *Mar. Biol.* **9,** 183.

Walker, J. D., and Colwell, R. R. (1974).   Microbial degradation of model petroleum at low temperatures.   *Microb. Ecol.* **1,** 63.

Wilson, R. D., Monaghan, P. H., Osanik, A., Price, L. C., and Rogers, M. A. (1974).   Natural marine oil seepage.   *Science,* **184**(4139), 857.

Chapter 5

Nutrients, Productivity, and Eutrophication

I. INTRODUCTION

The presence of inorganic nutrients in water and the resulting increase in plant productivity has become a serious water quality consideration. Clear, low-productivity waters can be rapidly changed into turbid, often odorous waters with wildly fluctuating oxygen levels when overenriched with plant nutrients. The large variation in oxygen level that is witnessed in nutrient-enriched waters is largely owing to excessive plant growth. This plant growth is usually in the form of algae or large-rooted hydrophytes that grow on the bottoms of streams and lakes. Often the nutrient-enrichment process changes the species diversity of a water system, resulting in large populations of a single species. Minerals such as nitrogen and phosphorus, which in themselves appear to be harmless, are therefore capable of causing environmental perturbations by accelerating plant productivity.

The process by which bodies of water become enriched, and hence more productive, is termed "eutrophication." Limnologists use the terms eutrophic, mesotrophic, and oligotrophic to describe the productivity state of a body of water. These terms were originally introduced by Weber (1907) to describe qualitatively the nutrient situation in German bogs. Today engineers, ecologists, and limnologists use these same terms to describe productivity states related to the nutrient enrichment of natural waters. There is no precise way to quantify each of these terms so that any body of water can be classified as either eutrophic, mesotrophic, or oligotrophic. The terms as originally proposed by Weber refer to relative states of organic matter production in German peat bogs. Recently, researchers have attempted to classify the productivity states of various bodies of water around the world. In some instances concentrations or loading rates of the nutrients nitrogen and phosphorus have been used; in other cases the rate of organic matter production in the system has been used. In all cases the systems cannot be correlated easily with one another; thus, each of these classifications should be used only in its general sense, or as it reflects the trend of a body of water going from a given productivity state to a higher productivity state. A general definition of each of these terms follows.

(1) *Oligotrophic* indicates a nonproductive water, is associated with low biological activity, and is usually related to geologically young bodies of water. Water quality is excellent in this case.

(2) *Mesotrophic* indicates a water with average productivity and is associated with some biological activity, but is still a balanced system. Water quality is good in this case.

(3) *Eutrophic* indicates a highly productive water. Excessive biological activity causes large fluctuations in environmental parameters. Water quality is usually significantly degraded.

From these definitions the reader can easily see that the terms eutrophic, mesotrophic, and oligotrophic should only be used to describe the progress of a system from a low-productivity state to a high-productivity state. We shall see in this chapter how the addition of various nutrients, especially nitrogen and phosphorus, to water systems causes a shift toward the eutrophic state.

The eutrophication process is essentially initiated by addition of the macronutrients nitrogen and phosphorus to water systems. Inputs of these nutrients come from domestic, industrial, and agricultural waste. Domestic sewage, even though treated at a secondary level, still contains large concentrations of nitrogen and phosphorus. Agricultural waste can contain large concentrations of nutrients, especially nitrogen. Treatment procedures have been initiated for nutrient removal from both agricultural waste and domestic sewage, and these procedures and their resultant effects on water quality will be discussed in Chapter 10.

It has also been shown that stormwater runoff from urban land has a large potential for contaminating surface waters with nutrients. Many systems, both lakes and rivers, that are bordered by highly populated areas exhibit high productivity even though there are no direct inputs of domestic sewage or industrial waste. Nitrogen and phosphorus is washed from lawns and streets during storms and transported directly to water systems. The resultant eutrophication of these waters is identical to that for a system that has been enriched through direct inputs of domestic sewage.

The nutrients required in large quantities for algal growth, that is, carbon, nitrogen, phosphorus, sulphur, and iron, are absorbed by plants in a fixed stoichiometric ratio. This ratio is relatively constant and can be used to make predictions of the trophic state and subsequent productivity potential of a given water system. We shall see that the concept of limiting nutrients, that is, when one of the nutrients is present in such low concentration that production is limited, is the basis for many methods of productivity control. An example of this would be a small pond in which phosphorus has been removed through chemical precipitation to such an extent that algal populations decline even though there are excessive amounts of other nutrients in the system. This law of limiting nutrients was first proposed by Liebig, and we shall examine Liebig's law and its application to water quality in Section VII.A.

To understand human impact on the eutrophication process and possible engineering alternatives for its control, the reader must first understand natural nutrient cycles. Each of the macronutrients required for algal growth undergoes natural cycling. Carbon, nitrogen, phosphorus, sulfur, and iron are cycled constantly between the inorganic and organic state, and it is this natural cycling that is the basis for all life on earth.

II. THE CARBON CYCLE

Perhaps the most important element in all biological units, at least in terms of concentration, is carbon. We shall assume that the reader understands basic physical chemistry and can follow the stoichiometry that will be presented. The chemistry of the carbon system, that is, carbon dioxide dissolved in aqueous systems, is very complex. The reader is referred to the text by Snoeyinck and Jenkins (1981) for an in-depth review of the carbonate–carbon equilibrium system. This section will outline the basic concepts of carbon dioxide in aqueous systems and the role of carbon dioxide as the inorganic carbon nutrient taken up by plants during their growth processes.

A. Inorganic Carbon and Alkalinity

Carbon dioxide present in the atmosphere readily dissolves into aqueous systems. Atmospheric carbon dioxide is present at concentrations of approximately 400 ml l^{-1}, and at equilibrium conditions approximately 0.7 ml l^{-1} of free carbon dioxide can be dissolved in water. When free carbon dioxide is dissolved in water, it readily disassociates and sets up an equilibrium system consisting of free carbon dioxide, bicarbonate ion, and carbonate ion. The relative percentage of each of the carbon species present at any time is a function of pH. The following equations reflect the disassociation process:

$$CO_2 + H_2O \rightarrow H_2CO_3; \tag{1}$$
$$H_2CO_3 \rightarrow HCO_3^- + H^+, \quad k_1 = 10^{-6.3}; \tag{2}$$
$$HCO_3^- \rightarrow CO_3^{2-} + H^+, \quad k_2 = 10^{-10.25}. \tag{3}$$

In addition, we can calculate the ionization fractions of each of the components of the carbonate system. This, as we shall see later, is useful in calculating the total carbon concentration that is present in a given water system at any time. The total carbon concentration is

$$[C_T] = [H_2CO_3] + [HCO_3^-] + [CO_3^{2-}]. \tag{4}$$

The ionization fractions are

$$[H_2CO_3] = [C_T]\alpha_0, \tag{5}$$
$$[HCO_3^-] = [C_T]\alpha_1, \tag{6}$$
$$[CO_3^{2-}] = [C_T]\alpha_2, \tag{7}$$

where

$$\alpha_0 = \left(\frac{1 + k_1}{[H^+]} + \frac{k_1 k_2}{[H^+]^2} \right)^{-1}, \tag{8}$$

$$\alpha_1 = \left(\frac{[H^+] + 1}{k_1} + \frac{k_2}{[H^+]} \right)^{-1}, \tag{9}$$

$$\alpha_2 = \left(\frac{[H^+]^2}{k_1 k_2} + \frac{[H^+]}{k_2} + 1 \right)^{-1}. \tag{10}$$

As seen in Eqs. (1)–(3), the disassociation of carbon dioxide or carbonic acid is dependent on pH, and, as mentioned previously, the relative distribution of each carbon species is also a function of pH. Figure 5.1 shows the relative distribution of each carbon species as a function of pH. Note that for low pH values, that is, less than 6.2, carbon dioxide (or H_2CO_3) is predominant. In the neutral range, pH values from 6.2 to 10.3, the bicarbonate ion HCO_3^- is the predominant form. For pH values greater

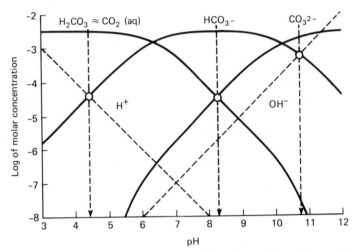

FIGURE 5.1   THE DISTRIBUTION OF CARBON SPECIES AS A FUNCTION OF pH.

than 10.3, the carbonate ion CO_3^{2-} is predominant. This distribution of inorganic carbon in water systems is important when considering the resultant plant growth potential.

Atmospheric carbon dioxide that is in equilibrium with water surface is constantly being dissolved into the aqueous phase. Once the carbon dioxide is dissolved, it forms carbonic acid, a weak acid, which then disassociates into bicarbonate or carbonate ion depending on the pH of the water. The relative distribution of free carbon dioxide, bicarbonate ion, and carbonate ion is then a function of pH. Once the total inorganic carbon concentration of a water system is known, one can determine the relative percentage of all forms of inorganic carbon present by using Fig. 5.1.

The most common parameter in use today for measuring the total inorganic carbon of a system is "alkalinity." Alkalinity is essentially a measure of the ability of a water system to resist a pH change when an acid is added. In general, alkalinity is a summation of all the components in a water system that act to buffer the water against the pH change. The general alkalinity relationship is

$$\text{alkalinity} = [HCO_3] + 2[CO_3^-] + [B(OH)_4^-] + [NH_3] + [OH^-] - [H^+]. \tag{11}$$

Equation (11) shows the components that affect the alkalinity of a system in which borate ion and ammonia are present. In natural water systems the concentrations of cations is small compared to the concentration of the carbonate anions; thus, the alkalinity of the carbonate-dominated system is simplified to

$$\text{alkalinity} = [HCO_3^-] + 2[CO_3^{2-}] + [OH^-] - [H^+]. \qquad (12)$$

It can readily be seen that substitution of equations (6) and (7) into equation (12) yields a relationship that relates alkalinity to total carbon, that is,

$$\text{alkalinity} = [C_T](\alpha_1 + 2\alpha_2) + [OH^-] - [H^+]. \qquad (13)$$

Thus, by measuring the alkalinity and pH of a water sample, it is possible to determine the total carbon present in the system. This parameter is important when attempting to predict the eutrophication potential of a water system. In addition, it helps us to analyze water systems for the limiting nutrient, or that nutrient that controls algal productivity.

It is important at this point to emphasize that these calculations reflect equilibrium conditions. This means that Eq. (13) would only be applicable if the system had time to equilibrate. This seldom occurs in natural systems. In reality, the rate that carbon dioxide is dissolved in water from the atmosphere is slow compared to the rate that it can be taken up by plants. Thus, we shall see that plants probably use carbon from the alkalinity pool for their growth instead of free carbon dioxide from the atmosphere.

B. Carbon Uptake by Plants

The fixation of inorganic carbon by plants is the first step in converting inorganic carbon to organic carbon. This part of the cycle occurs during all daylight hours by photosynthesis in algae and rooted plants. It is not clear whether algae and other plant types consume free aqueous CO_2 or bicarbonate ion. From a water quality perspective it makes little difference, because we are concerned with the total carbon available for plant growth. We should recall that the concentrations of the carbon forms are interdependent, so that the withdrawal of any one form affects the other species. We should also recall that this distribution is pH dependent. Thus, if the plants are using the bicarbonate ion as their carbon source, carbon from the carbonate and carbonic acid forms will change to bicarbonate as it is taken up by the plants.

Photosynthesis only proceeds during daylight hours. In the darkness, the plants respire, that is, take up oxygen and release CO_2. Hence, carbon dioxide is returned to the dissolved carbon pool. This "short-circuiting" of the carbon cycle can be a major factor, depending on the biomass of a given system. In general, however, we shall consider the carbon fixed by photosynthesis to be organic carbon, and hence available for consumption by herbivores or microorganisms.

Carbon is incorporated into plant cells in a myriad of forms, including sugars and proteins. From an engineering point of view, however, this is not of interest, for we are only concerned with the percentage of total carbon tied up in organic fractions. Thus, we measure carbon in the organic form by the well-known BOD, COD, or total carbon analyses. When plants die, bacteria are capable of degrading the plant material, releasing CO_2 in the process. CO_2 is then returned to the system as inorganic carbon. The process uses dissolved oxygen to oxidize the carbon compounds, and thus oxygen is removed from the system.

In summary, inorganic carbon is introduced to the cycle through equilibrium with the atmosphere, used by plants in photosynthesis, coverted to organic carbon, mineralized by bacteria upon death, and returned to the cycle as CO_2. Figure 5.2 shows a typical carbon cycle, in which inorganic carbon is converted to organic carbon and returned through mineralization to CO_2. As mentioned previously, the carbon cycle is intimately related to the oxygen balance of the system, and the cycle shown in Fig. 5.2 reflects the oxygen requirement during the mineralization process. It is this requirement for oxygen that depletes the oxygen reserves in natural systems when the organic carbon pool becomes too large.

C. Water Quality Aspects

Inputs from human activity can alter the carbon cycle in many ways. Excessive inputs of organic carbon, that is, sewage and/or industrial waste,

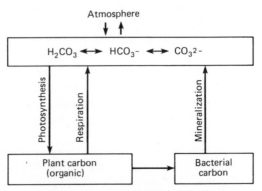

FIGURE 5.2   THE CARBON CYCLE.

can create a large oxygen demand in receiving waters and in severe cases purge the system of oxygen entirely. This obviously upsets the carbon cycle, as well as the dissolved oxygen balance. Also, increasing other inorganic nutrients causes increased algal growth and subsequently a large accumulation of organic matter, once again upsetting the carbon cycle.

Another important parameter that affects water quality is the pH – carbon dioxide relationship, which is perturbed by excess algal growth. In a system that is overly enriched with nutrients and in which excessive algal growths are present, large amounts of carbon dioxide will be used during photosynthesis. During the daytime hours when CO_2 is being consumed, the pH will rise. In eutrophic situations daytime pH values as high as 9 or 10 are common. When the pH is elevated to these levels, other physiological processes are affected. As an example, many elements that are common in natural water systems become toxic to animals and plants when the pH is increased. In addition, calcium carbonate tends to precipitate out of solution during periods of high pH. In lakes, for example, this precipitation increases the sediment load.

It is instructive to note that the concentration of CO_2 does not enter into our alkalinity equations. Recall that CO_2 only affected the total carbon budget; therefore, as CO_2 is removed during photosynthesis and the pH is elevated, we should expect very little change in the alkalinity of the water. The alkalinity changes, however, if the pH is increased sufficiently that calcium carbonate precipitates out of the system. A eutrophic lake has a widely fluctuating pH value, with subsequent fluctuations in the concentration of dissolved oxygen. Unless calcium carbonate is precipitated from the system, however, there will be no change in alkalinity. This wide variation in the concentration of carbon dioxide, with subsequent pH shifts, can degrade water quality and impair the use of the water for its intended purpose.

III. THE NITROGEN CYCLE

Molecular nitrogen, one of the most common elements in the atmosphere, dissolves readily into aqueous systems. Molecular nitrogen is biologically inert and does not enter into very many biological processes. One of the reasons for this relative inertness is the N – N triple bond, which is extremely strong and requires over 220 kcal mole^{-1} of energy to break. There is a relatively small class of organisms that are capable of using molecular nitrogen as a nitrogen source. These are called nitrogen fixers, and they will be discussed in Section III.A.

When molecular nitrogen is dissolved in aqueous systems, it is eventually converted to aqueous forms, predominately ammonia (NH_3), nitrite ion (NO_2^-), and nitrate ion (NO_3^-). These forms of nitrogen, with the exception of NO_2^-, are readily available for biological activity and enter into most growth processes. The oxidation state of nitrogen in these compounds varies from -3 to $+5$, and Table 5.1 lists the common forms of nitrogen according to valance and order of abundance. Note that molecular nitrogen is the most abundant form in water systems. Of the aqueous forms of nitrogen, nitrate is generally the most common, with ammonia being second and nitrite being third. It should be emphasized that this order of abundance is representative of a natural system, and when humans perturb the system one of the other forms of nitrogen may become dominant.

All of the forms of nitrogen in Table 5.1 can be present in environmental situations, because nitrogen can be easily oxidized and/or reduced by many environmental processes. It should be mentioned here that we have listed ammonia as the gas (NH_3), but it can also appear as ammonium ion (NH_4^+). The relative distribution of NH_3 and NH_4^+ is dependent on the pH of the water, and diagrams very similar to those constructed for the carbonate system can be constructed for the ammonia–ammonium equilibrium. Such a diagram would show that the equilibrium pH for the ammonia–ammonium system is approximately 9.3; therefore, in natural systems of neutral pH, ammonium ion is the predominant form. For systems in which the pH has been increased, such as high-productivity lakes, a significant amount of ammonia gas volatilizes to the atmosphere. This will be discussed later, and we shall see that it is one of several mechanisms for nitrogen removal in aquatic systems.

A. Nitrogen Uptake by Plants

In the beginning of this chapter it was mentioned that nitrogen gas or elemental nitrogen is fixed in aquatic systems and converted to various

TABLE 5.1   INORGANIC SPECIES OF NITROGEN IN DECREASING ORDER OF ABUNDANCE

SPECIES	SYMBOL	VALENCE
Molecular nitrogen	N_2	0
Nitrate	NO_3^-	$+5$
Ammonia	NH_3	-3
Nitrite	NO_2^-	$+3$

nitrogen compounds. It was also noted that very few organisms were capable of using N_2 directly as a nitrogen source, owing to its strong $N-N$ triple bond. Nitrogen is fixed by many bacteria and several strains of blue-green algae. By the term "fixed" we refer to the fact that nitrogen gas is used as a nitrogen source directly by the organisms and converted to cellular organic nitrogen.

The main input of new nitrogen to a system usually comes from nitrogen fixation. Other forms of nitrogen tend to be cycled in the system or lost through gasification. In some systems the amount of N_2 fixed by organisms is large, and substantial amounts of nitrogen can be added to the cycle on a daily basis. Common agricultural practice relies on symbiotic bacteria within legumes to fix nitrogen in the soil. Crops are rotated on a yearly basis, with nitrogen fixers being grown to replenish nitrogen that has leached from the fields. The following list shows some of the more common forms of microflora that are known to fix nitrogen (Alexander, 1961).

(1) Heterotrophic bacteria: *Achromobacter, Aerobacter, Azotobacter, Azotomonas, Bacillus polymyxa, Beijerinckia, Clostridium, Psuedomonas.*
(2) Chemoautotrophic bacteria: *Methanobacillus omelianskii.*
(3) Blue-green algae: *Anabaena, Anabaenopsis, Aulosira, Calothrix, Cylindrospermum, Nostoc, Tolypothrix.*
(4) Photosynthetic bacteria: *Chlorobium, Chromatium, Rhodomicrobium, Rhodopseudomonas, Rhodospirillum.*

Several environmental factors are known to affect the rate of nitrogen fixation in a natural system. Certainly temperature and pH are factors that control the rate to a great extent. However, more important is the fact that other nitrogen compounds, especially ammonia, regulate the ability of organisms to use molecular nitrogen as a nitrogen source. If ammonia is added to a system, nitrogen fixation is inhibited because organisms prefer to use ammonia for a nitrogen source rather than fix atmospheric nitrogen. Cells that are fixing nitrogen can readily shift to using ammonia for their source of nitrogen without any lag in growth.

One of the main pathways of nitrogen utilization in natural systems is photosynthesis. Algae and rooted plants use different nitrogen compounds during their growth cycle. Normally ammonia and nitrate are favored over molecular nitrogen as a nitrogen source. When a plant uses nitrogen during its metabolism, the nitrogen must be reduced to ammonia to be assimilated into the plant cell. Thus, if a plant is using nitrate, the nitrogen must be reduced by an enzyme before being taken up by the plant. This costs the plant energy, and in most cases means that the plant would require or prefer to use ammonia. In this case the plant has to expend very little energy to reduce the nitrogen, and it is mostly for this reason that plants often take ammonia out of solution preferentially to nitrate. The enzyme required for reducing nitrate (nitrate reductase) is suppressed by the presence of ammonia.

The prediction of algal growth response as a function of nitrogen concentration has been attempted by the Michaelis–Menten enzyme kinetics model as follows:

$$\mu = \mu_{max}S/(K_s + S), \tag{14}$$

where μ is the growth rate of the organism, S the ambient concentration of the nutrient, and K_s a half-saturation constant.

Using this rational function it is possible to calculate half-saturation constants K_s for various algal forms as a function of nutrient type. It is therefore possible to compare the relative growth capacities of various plants for various nutrient situations. This is very useful from an engineering point of view in that it allows one to predict which algal form should be present for a given nutrient situation. This knowledge, of course, is imperative when we recall that certain forms of algae are more of a water quality problem than others. It should be mentioned that the Michaelis–Menten equation is intended to model only one nutrient that has been designated as the limiting nutrient. If, for instance, we use Eq. (14) for nitrate, then we assume that phosphorus, sulphur, iron, and other elements are all present in excess, that is, only nitrate is regulating the growth of the organism. In addition, if nitrate is used in the model, then ammonia must not be present in any appreciable concentration, so that the enzyme system for nitrate reductase is not hindered.

Figure 5.3 shows some typical growth curves for various species of algae. It can be seen that each curve is unique, the growth rate depending on the

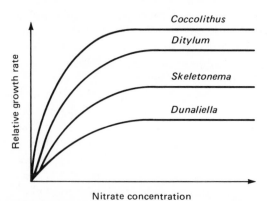

Nitrate concentration

FIGURE 5.3   THE EFFECT OF NITRATE CONCENTRATION ON THE GROWTH OF FOUR ALGAE AT TWO DIFFERENT LIGHT LEVELS. Completely different responses to nitrate are observed for each algae at different light intensities. (From Eppley, 1969.)

ability of the algae to use nitrate. Calculated K_s values for each of the curves would also differ, and hence affort a measuring stick for algal response.

Once inorganic nitrogen, either in the form of ammonia or nitrate, has been assimilated, it is converted to organic nitrogen. Organic nitrogen is usually written as NH_2, with the oxidation state of nitrogen similar to that of ammonia. Many organic compounds in the cell contain nitrogen, but they are not of interest from a water quality perspective. The main organic forms, however, are urea, uric acid, and various amines. These compounds are released either by leaching from the plant cell or when the plant cell lyses. The cycle is now complete, because the organic fraction of nitrogen is returned to the water system. The organic forms of nitrogen are readily mineralized to ammonia, which then follows the cycle back to nitrite and nitrate.

It has also been shown that in certain cases organic nitrogen compounds can be used directly by plants for their nitrogen metabolism. This appears as a small shunt in our nitrogen cycle diagram (see Fig. 5.4). It is of interest to know this occurs, but evidence to date does not indicate that a large amount of organic nitrogen follows this pathway. In general, inorganic nitrogen is a better and much-preferred source of nitrogen for plants than is organic nitrogen.

B. Nitrification and Denitrification

Ammonia that enters water systems either from decaying plant material or sewage discharges is readily oxidized to nitrite and nitrate if sufficient oxygen is present. This process is mediated by two forms of bacteria that essentially gain their energy from the oxidation of ammonia and nitrite. The two forms of bacteria are called *Nitrosomonas* and *Nitrobacter,* and they are present in almost all aquatic systems. These bacteria are usually found together, because *Nitrosomonas* is only capable of mediating the oxidation of ammonia to nitrite and *Nitrobacter* can only mediate the oxidation of nitrite to nitrate. The following equation shows the oxidation of ammonia to nitrite:

$$NH_4^+ + OH^- + 1.5O_2 \rightarrow H^+ + NO_2^- + 2H_2O + 54.9 \quad \text{kcal.} \tag{15}$$

According to this equation the process is pH-dependent, with the optimum pH for rapid nitrification of ammonia being about 8.0 to 8.5.

After the oxidation of ammonia has been mediated by *Nitrosomonas,* *Nitrobacter* then mediates the oxidation of nitrite to nitrate according to the following equation:

$$NO_2^- + 0.5O_2 \rightarrow NO_3^- + 18 \quad \text{kcal.} \tag{16}$$

Note from reactions (15) and (16) that a total of 4.57 g O_2 g^{-1} NH$_4^+$ is required for the oxidation. Therefore, the nitrification of ammonia has the potential for removing large amounts of oxygen from the system.

Both *Nitrosomonas* and *Nitrobacter* are strict chemoautotrophs, meaning that they use inorganic carbon as their carbon source. In fact, organic carbon inhibits their growth; thus, they do not grow well in systems containing large amounts of organic carbon. This is why in sewage treatment plants a large portion of the organic carbon or BOD must be removed from the system before the nitrifying bacteria can initiate the nitrification process.

Denitrification is the process by which nitrate is reduced to nitrite and then to nitrogen gas. Once again, this process is microbially mediated in the environment and is common to many forms of bacteria. In anaerobic (low-oxygen) systems many forms of bacteria utilize bound oxygen in forms such as NO$_3^-$. As the oxygen concentration in a system is lowered (as the system becomes anaerobic), bacteria start using oxygen bound in nitrate, sulphate, and other compounds. When the nitrate ion is used, for example, the nitrogen is reduced from nitrate to nitrite as the oxygen is removed by the bacteria. In most cases it is further reduced to nitrogen gas, which then leaves the system. It is by denitrification, then, that nitrogen is lost from the nitrogen cycle. As a matter of fact, it has been postulated that this mechanism accounts for the majority of the nitrogen lost from natural cycles. Once again, it should be emphasized that the denitrification process is common to many groups of bacteria, whereas the nitrification process is limited to a few chemoautotrophs.

We have now described the main processes involved in the natural nitrogen cycle. Atmospheric nitrogen is fixed by a specific group of organisms into ammonia, nitrate, and nitrite. This nitrogen then becomes available for general biological use and is taken up by plants in their growth cycles. Both ammonia and nitrate can be utilized by plants as a nitrogen source. During growth the oxidized forms are reduced to organic nitrogen compounds, and these nitrogen compounds are then returned to the aqueous system either by the death of the plants or natural leaching. Organic nitrogen is readily converted to ammonia, which is then either reused by plants or oxidized to nitrate in reactions mediated by *Nitrosomonas* and *Nitrobacter*. Nitrate, the most oxidized form of nitrogen, can either be reused by plants or reduced to nitrogen gas by the denitrification process, which proceeds in low-oxygen situations and is mediated by many forms of bacteria. The main nitrogen input to the cycle in natural systems is the fixation of atmospheric nitrogen, and the main output is the loss of nitrogen gas to the atmosphere from denitrification. When a natural system is perturbed, other entrances and exits to the cycle occur. Here, we should mention that unlike other cycles, e.g., the phosphorus cycle, very little loss of nitrogen to the sediments occurs by inorganic precipitation. Very few inorganic compounds are formed with nitrate or ammonia, and hence precipitation does not appear to be an important pathway in the loss of nitrogen from the cycle. Figure 5.4 shows a schematic diagram of the nitrogen cycle.

C. Water Quality Aspects

The natural nitrogen cycle can be upset, as can other cycles, by human interference. Discharges of various forms of nitrogen to water systems by human activity tend to make the cycle lopsided and cause environmental perturbations. Examples of this are inputs of domestic sewage and industrial waste to water systems. Municipal sewage can upset the nitrogen cycle because of large inputs of ammonia.

According to Eqs. (15) and (16), the nitrification of discharged ammonia requires large amounts of oxygen which, depending on the situation, can deplete the system of its oxygen reserves. Often the added demand for oxygen causes a depression of the oxygen level below that required for complete carbon stabilization. We also recall that ammonia becomes a readily available source of nitrogen for plant growth, and regardless of whether the ammonia is oxidized to nitrate or not, it can readily be used by plants. Thus, by discharging wastes containing ammonia, eutrophication will be stimulated, assuming that excess nitrogen is required for growth in the system.

When plant material, either in algal or rooted-plant form, takes up ammonia from the water, it forms material that eventually dies, and the

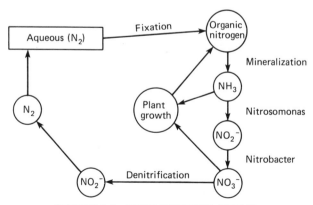

FIGURE 5.4   THE NITROGEN CYCLE.

ammonia then becomes mineralized and released back into the system. If significant amounts of industrial or domestic wastes have been discharged into the system, a tremendous amount of plant material is generated in addition to the ammonia discharge, which also contributes to the cycle. Thus, such discharges increase the amount of nitrogen that is going around in the nitrogen cycle, and hence increase the overall productivity of the system. The final result of waste discharges to a lake or river system is a layer of decaying plant material in the sediments. We must then be concerned with the rate at which nitrogen becomes mineralized from this bottom layer back into the system in addition to the rate at which nitrogen is being put into the system through domestic waste discharges. In severe cases, even if the discharge of nitrogen from domestic waste is halted, the input of nitrogen from the sediments can be sufficient to maintain a high level of productivity in the system.

In an aqueous system that has had a substantial nitrogen addition, there is an increase in productivity and hence a larger resultant biomass of plant material. In such cases, as discussed for the carbon system, a fluctuation in pH as well as fluctuations in dissolved oxygen concentration can develop. As the pH is increased due to photosynthetic activity, ammonium ion is converted to ammonia gas, and large amounts of nitrogen are removed from the system. In these situations a new exit from the nitrogen cycle has been developed. Many severely perturbed systems have shown this rate of ammonia loss to be large, and in some instances it represents the majority of the nitrogen lost from the cycle.

IV. THE PHOSPHORUS CYCLE

A. Forms of Phosphorus

The element influencing the eutrophication process that has received the most attention is phosphorus. Listed as the 11th most common element in igneous minerals in the earth's crust, phosphorus is usually found as orthophosphate, and the most common phosphorus-containing mineral is apatite. A few members of the apatite group are hydroxyl apatite, $Ca_{10}(OH)_2(PO_4)_6$; fluoroapatite, $Ca_{10}F_2(PO_4)_6$; and carbonate fluoroapatite, $(CaH_2O)_{10}(FOH)_2(PO_4CO_3)_6$. In addition to these minerals, phosphate is also found adsorbed onto clay surfaces and combined as metal phosphates. Actually, we shall see that phosphorus forms many inorganic precipitates and can be readily transported out of the phosphorus cycle. This is different from the nitrogen and carbon cycles, in which inorganic precipitates are not an important consideration.

Phosphorus can also be found in particulate organic fractions suspended in the water column. This particulate matter is organically bound phosphorus that has been taken up by microflora or bacteria. Stumm and Morgan (1981) have classified the solid phases of phosphorus commonly found in nature as follows:

(1) soil and rock mineral phases,
(2) mixed phases,
(3) suspended or insoluble organic phosphorus.

A few examples of each of these types of solid phosphorus are given in Table 5.2.

One of the main differences in the phosphorus cycle in comparison to the nitrogen cycle is the fact that phosphorus occurs in nature almost exclusively as phosphate (PO_4^{3-}). Phosphate is dissolved in aqueous systems owing to the weathering of rocks and, once dissolved, orthophosphate can occur in any one of three forms, depending on the pH. Figure 5.5 shows the effect of

TABLE 5.2 SOLID PHASES OF PHOSPHORUS OF POTENTIAL SIGNIFICANCE IN NATURAL WATER SYSTEMS[a]

FORM	REPRESENTATIVE COMPOUNDS OR SUBSTANCES
Soil and rock mineral phases	
Hydroxylapatite	$Ca_{10}(OH)_2(PO_4)_6$
Brushite	$CaHPO_4 \cdot 2H_2O$
Carbonate fluorapatite	$(Ca,H_2O)_{10}(F,OH)_2(PO_4,CO_3)_6$
Variscite	$AlPO_4 \cdot 2H_2O$
Strengite	$FePO_4 \cdot 2H_2O$
Wavellite	$Al_3(OH)_3(PO_4)_2$
Mixed phases, solid solutions, sorbed species, etc.	
Clay–phosphate (e.g., kaolinite)	$[Si_2O_5Al_2(OH)_4 \cdot (PO_4)]$
Metal hydroxide–phosphate	$[Fe(OH)_x(PO_4)_{1-x/3}], [Al(OH)_x(PO_4)_{1-x/3}]$
Clay–organophosphate	$[Si_2O_5Al_2(OH)_4 \cdot ROP]$, clay–pesticide, etc.
Metal hydroxide–inositol phosphate	$[Fe(OH)_3 \cdot$ inositol hexaphosphate]
Suspended or insoluble organic phosphorus in bacterial cell material, plankton material, plant debris, and proteins	Inositol hexaphosphate or phytin, phospholipid, phosphoprotein, nucleic acids, polysaccharide phosphate

[a] From Stumm and Morgan (1981). © 1981 John Wiley & Sons, Inc.

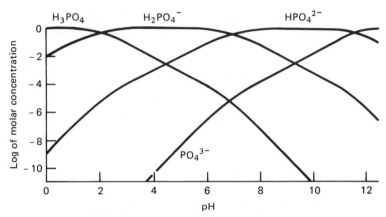

FIGURE 5.5   THE DISTRIBUTION OF PHOSPHORUS SPECIES AS A FUNCTION OF pH. (From Stumm and Morgan, 1981. © 1981 John Wiley & Sons, Inc.)

pH on various forms of phosphorus. In the netural pH range, the predominant forms are $H_2PO_4^-$ and HPO_4^{2-}. All of these species are designated as orthophosphate and will be considered to be available for uptake by plants. In addition, it should be mentioned that phosphorus does not undergo a change in oxidation state, as occurred with nitrogen.

Phosphates readily complex with cations in water to form insoluble compounds. In particular, phosphate readily complexes with iron, aluminum, and calcium in natural systems. The occurrence of phosphate in any system, then, is governed in part by the solubility of the various metal phosphate compounds that are formed. Figure 5.6 shows that the relative solubility, or the ability of these compounds to precipitate out of solution, is once again primarily a function of pH. For natural waters (pH ≈ 7), iron and aluminum phosphate compounds appear to be the least soluble and therefore govern the system. At pH ≈ 8 or greater, the calcium phosphate compounds exhibit the lowest solubility, and would be expected to govern the system in this pH regime.

The formation of metal phosphate compounds is also dependent on the oxidation–reduction potential of the system. As we have seen previously, iron phosphate compounds in particular are strongly influenced by environmental shifts in the oxidation–reduction potential. In the oxidized form Fe^{3+}, iron phosphates are very insoluble and tend to precipitate out of the system. Once reduced to Fe^{2+}, however, iron and phosphorus are redissolved and returned to the system. This is one of the problems with allowing low oxygen levels in lakes and rivers.

The resultant form of phosphate that occurs in water systems is important when predicting the environmental effects of phosphorus discharges.

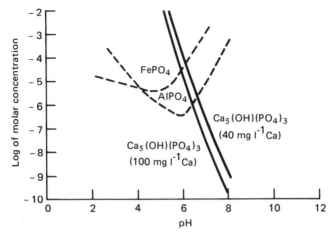

FIGURE 5.6   THE SOLUBILITY OF SELECTED METAL PHOSPHATE COM-
POUNDS AS A FUNCTION OF pH.   (From Stumm and Morgan, 1981. © 1981
John Wiley & Sons, Inc.)

Leckie and Stumm (1970) presented data that reflect the amount of orthophosphate as a percentage of the total phosphorus for various systems.
Table 5.3 shows these values, and it should be noted that orthophosphate
represents a relatively small portion of the total phosphorus content.   This is
owing to increased algal growth with the subsequent uptake of orthophosphate, especially in eutrophic lakes.   Note that effluents from secondary

TABLE 5.3   ORTHOPHOSPHATE AND TOTAL PHOSPHORUS
CONCENTRATIONS FOR VARIOUS WATER SYSTEMS[a]

SYSTEM	TOTAL PHOSPHORUS (mg/l)	% PHOSPHORUS PRESENT AS SOLUBLE ORTHOPHOSPHATE
Domestic wastewater	5–20	15–35
Effluents from secondary treatment plants	3–10	50–90
Agricultural drainage	0.05–1.0	15–50
Unpolluted lakes	0.01–0.04	10–30
Eutrophic lakes	0.03–1.5	5–20
U. S. rivers	0.01–1.0	—
Oceans (mean value)	0.07	—
Rainwater	0.004–0.03	—

[a] From Leckie and Stumm (1970).

treatment plants contain high percentages of orthophosphate, owing to the hydrolization of condensed polyphosphates by the treatment process. These polyphosphates are easily broken down by microorganisms in the sewage and leave the treatment plant as free or available orthophosphate.

B. Phosphorus Uptake by Plants

As was the case for other nutrients, phosphorus is required for the growth of plants and bacteria.   Once assimilated, orthophosphate is converted to organic phosphorus.   In this case the phosphorus is converted to polyphosphate, that is, a chain of phosphates sometimes referred to as tripolyphosphate or pyrophosphate.   The general term for these phosphorus compounds is condensed phosphates, and the phosphorus in plant and bacterial cells that we consider to be organic phosphorus is in the form of these condensed phosphates.   Once again, when the plant cell lyses these condensed phosphates are discharged back into the aqueous system.   This condensed form of phosphate is not readily available for plant uptake and must be hydrolized by bacteria to orthophosphate before it can be used by plants.   Thus, the organic part of the phosphorus cycle consists of the uptake of orthophosphate, the conversion to condensed phosphate in plant cells, the release as polyphosphate, and the hydrolization by bacteria back to orthophosphate.

The fact that polyphosphates cannot be used directly by plant cells for their phosphorus requirement is important in environmental systems. Detergents contain polyphosphate builders that are discharged by sewage systems, but these condensed phosphorus compounds are not available for plant uptake.   However, once they pass through a secondary treatment process, they can be readily hydrolized by microorganisms in the sewage to orthophosphate.   Thus, effluents from secondary treatment plants can be major sources of usable phosphorus.

C. Equilibria of the Phosphate System

We can now piece together the phosphorus cycle, which contains both biological, physical, and chemical interactions.   Orthophosphate that becomes dissolved owing to the weathering of minerals is readily taken up by plants and converted to polyphosphate.   Upon the death of the plants, these condensed phosphates are released back into water systems and then rehydrolized by bacteria.   The resultant orthophosphate goes back into solution and can be recycled by plant uptake.   In addition, the pH and oxidation –

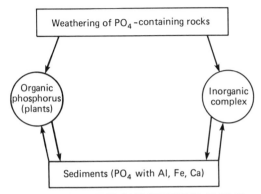

FIGURE 5.7   THE PHOSPHORUS CYCLE.

reduction potential in the water system affect the complexation of ortho-phosphate by various metal compounds. We have seen that iron, alumi-num, and calcium in natural systems form complexes with phosphate that precipitate out of the system. Thus, the phosphate cycle is a complex equilibrium of various phosphate phases. Note that solid inorganic phos-phorus is in equilibrium with soluble orthophosphate, which in turn is in equilibrium with soluble organic and particulate organic phosphorus.

Figure 5.7 shows the phosphorus cycle in a natural system. The main input of phosphate to the cycle is the weathering of minerals, and the main exit from the cycle is the precipitation of inorganic forms of phosphate.

D.   Water Quality Aspects

As for other cycles, the phosphorus cycle can also be upset by human activity. Domestic sewage inputs to aqueous systems, for instance, bring large amounts of phosphorus into the cycle. This increases plant productiv-ity and, as a result, eutrophication is accelerated. We have seen that the cycle has only one basic exit, the precipitation of inorganic phosphate. The environmental conditions regulating this process are variable, hence the biological uptake of phosphorus continues on its own cycle despite inor-ganic precipitation. Phosphate can be cycled within a system many times, contributing to increased biological activity before it is finally removed from the system by sedimentation.

In lakes that have a low oxygen tension in the hypolimnion, orthophos-phate can redissolve owing to a reduction in the concentration of the cationic compounds that have held it in an insoluble phase. Therefore, as for the nitrogen cycle, sediments should be evaluated as to their total phosphorus concentration and rate of phosphorus dissolution to the overlying waters. Also, in many situations bottom sediments contain enough phosphorus to accelerate eutrophication even after external sources have been terminated.

In many freshwater situations, phosphorus has been shown to be the limiting nutrient, that is, the nutrient in the lowest concentration relative to the stoichiometric ratio required for algal growth. Phosphorus can then become extremely influential in the growth potential of aquatic systems. Many workers have attempted to set limits on phosphorus concentrations that will prevent eutrophication. The original work by Sawyer (1947) on Wisconsin lakes has shown that a limit of 10 μg l^{-1} of phosphorus is the limit for preventing excessive phytoplankton growth. Other researchers have related phosphorus loading rates to the potential for eutrophication. At this point in time, it is not clear exactly which parameter is the most meaningful when attempting to predict the eutrophication potential of lake and river systems.

Perturbations in the phosphate cycle also directly affect other cycles. For instance, it can be shown that phosphorus discharges from sewage contribute to the creation of algal material, which becomes an oxygen demand in itself. Because of this, the oxygen demand in a system can be related directly to the amount of phosphorus that was added by the sewage. It is easy to see that by discharging phosphorus in sewage effluents oxygen can be totally consumed in receiving waters.

V. THE SULFUR CYCLE

Sulfur is also considered to be a macronutrient required for plant growth, owing to the large concentrations found in growing plant cells. Sulfur undergoes a natural cycle similar to that of nitrogen in that microbial processes dictate which form of sulfur appears in the natural environment. The most oxidized form of sulfur is sulfate, which is the form most readily assimilated by plants. From a water quality perspective, this can be considered the only form of sulfur that is incorporated into plant biomass. Once sulfate is taken up by plants, it is reduced to organic sulfur. In general, sulfur appears in plant and animal proteins, amino acids, and the B vitamins. Once plant cells die, bacteria are capable of turning this proteinaceous sulfur into reduced sulfide, or H_2S. In systems that contain very low oxygen concentrations, H_2S can build up and cause odor problems. Mercaptans can also be formed, further contributing to the deterioration of water quality. Once H_2S has been liberated to the water system, groups of specialized bacteria mediate the oxidation of H_2S to sulfate ion (SO_4^{2-}).

The oxidation states of sulfur found in the environment vary from -2 (sulfide) to $+6$ (sulfate). Table 5.4 shows the common forms of sulfur and their chemical formulas.

Because the sulfur cycle is controlled by bacterial activity, it is of interest here to mention the bacterial groups involved. The sulfur oxidizers, which are bacteria that mediate the oxidation of sulfur to sufate, can be divided into the following four groups:

(1)   nonfilamentous chemoautotrophic bacteria *(Thiobacillus),*
(2)   heterotrophic bacteria, fungi, and actinomycetes,
(3)   thread-forming bacteria *(Beggiatoa, Thiothrix),*
(4)   photosynthetic green and purple sulfur bacteria.

The most common sulfur oxidizers are group (1), and the sources of sulfur that can be used by this group are sulfur, sulfide, thiosulfate, tetrathionate, and trithionate. Most of the sulfur oxidizers are autotrophs and use CO_2 or bicarbonate as their carbon source. Few species are capable of oxidizing H_2S to elemental sulfur, the most common ones belonging to the genera *Beggiatoa* and *Thiothrix.* The purple and green sulfur bacteria are a specialized group that also oxidize H_2S to sulfur, but they are not present in large numbers in natural systems.

In summary, we have seen that the processes of mineralization or oxidation convert hydrogen sulfide to sulfate ion. These processes are catalyzed, or mediated, by several groups of bacteria. Each group is responsible for oxidizing a particular form to a higher valance state, which is similar to the nitrification process.

Once sulfur has been oxidized to sulfate ion, plants can assimilate it and reduce it to organic sulfur; also, several heterotrophic forms of bacteria are capable of reducing sulfate ion. In water systems that are low in oxygen, sulfate bacteria of the genus *Desulfovibrio* are able to reduce sulfate to H_2S

TABLE 5.4   CHEMICAL FORMULAS OF COMMON SULFUR FORMS

COMPOUND	FORMULA
Sulfate	SO_4^{2-}
Sulfite	SO_3^{2-}
Trithionate	$S_3O_6^{2-}$
Tetrathionate	$S_4O_6^{2-}$
Thiosulfate	$S_2O_3^{2-}$
Sulfur	S
Sulfide	S^{2-}
Bisulfide	HS^-
Hydrogen sulfide	H_2S

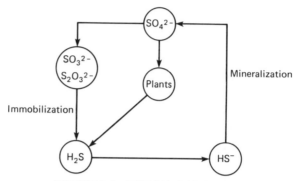

FIGURE 5.8   THE SULFUR CYCLE.

gas at a very rapid rate.   In fact, several species of *Desulfovibrio* are able to use sulfite, sulfate, thiosultate, tetrathionate, or colloidal sulfur and reduce them to H₂S gas.   Numerous genera of microorganisms are able to reduce the more-reduced forms of sulfur to H₂S, but it is only the *Desulfovibrio* group that is able to reduce the oxidized forms of sulfur.   Figure 5.8 shows the sulfur cycle as it has been described in this section.

We have now described the macronutrients required for plant growth, along with the main cycles that influence them.   Many other elements, for example, iron, manganese, and silica, are also required in lesser amounts for algal growth and in specific systems can in fact influence growth.   However, in most cases it is the macronutrients, i.e., carbon, nitrogen, phosphorus, and sulfur, that most often regulate algal growth, and it is these elements that are of interest from a water quality perspective.   Section VI will describe some of the minor constituents that have been shown to regulate algal growth.   Although they are not usually of significant interest in considerations of water quality, it is helpful that the reader understand the influence of these parameters on eutrophication.

VI.   MICRONUTRIENTS AND PRODUCTIVITY

Elements required for algal growth in small concentrations are often important considerations for water quality.   These constituents, including metals and organic growth factors, are required in small quantities and probably limit algal productivity only in isolated situations.   As mentioned earlier, macronutrients such as nitrogen, phosphorus, and sulfur tend to affect algal productivity on a large scale.   However, there are several published accounts of situations in which metals or organic growth factors limited algal productivity.   A few of these cases will be reviewed here so that the role of micronutrients in a water quality program can be better understood.

A. Trace Metals

Iron, manganese, silica, boron, molybdenum, zinc, cobalt, copper, vanadium, and several other elements are known to be required by algae for normal growth.   The concentration of these elements required is usually in the part-per-billion range, hence they are in sufficient supply in most water systems.   Goldman (1964) reported on cases in which the addition of certain metals to water samples stimulated higher productivity, and in these cases trace metals can be said to limit algal growth.   From the point of view of water quality, then, the addition of trace metals to a system can indeed influence eutrophication.   Most human-induced perturbations, however, are in the form of sewage discharges, stormwater runoff, or industrial waste inputs.   These inputs would more than likely also contain large amounts of one or more of the macronutrients, which would generally influence eutrophication to a greater extent.   We find, then, that the influence of trace metals appears to have a greater effect on productivity in relatively unproductive systems.   Examples of low-productivity systems would include open ocean waters and extremely oligotrophic lakes.

B.   Organic Growth Factors

Several complex organic compounds, especially vitamins, have been shown to be required by some algae for growth.   Provasoli (1969) compiled a list of algae grouped by vitamin requirements, which is shown in Table 5.5.   Most of the algal forms listed require vitamins or organic growth factors in the tenths to hundredths of a nanogram per liter range, which shows that extremely low levels of these organic compounds can saturate normal algal cells.   The effects of vitamins on algal growth are most pronounced in low-productivity regions such as the open ocean, because more fertile systems contain concentrations of organic growth factors sufficient to meet algal-growth requirements.

In natural systems bacteria are the main producers of vitamins, and approximately 70–80% of the soil bacteria in the rhizosphere produce vitamin B_{12}.   Therefore, common bacteria in the sediments or suspended particulate matter would probably produce enough vitamin B_{12} to prevent the limitation of algal growth.   We can see that once again the macronutrients will probably regulate the ability of algae to grow in a normal

TABLE 5.5 VITAMIN REQUIREMENTS OF SELECTED ALGAL GROUPS[a]

ALGAL GROUPS	NO VITAMINS	VITAMINS	B_{12}	THIAMINE	B_{12} AND THIAMINE	BIOTIN	RATIO OF B_{12} TO THIAMINE	MAJOR REQUIREMENT
Do not require biotin								
Blue-green algae	9	7	7	0	0	—	7/0	B_{12}
Chlorophytes	24	49	16	13	20	—	36/33	None
Diatoms	22	24	18	3	3	—	21/6	B_{12}
Cryptomonads	0	11	2	2	7	—	9/9	None
Require biotin								
Euglenids	0	11	—	1	9	1	—	None
Chrysomonads	1	27	1	9	14	5	17/24	Thiamine
Dinoflagellates	1	18	13	0	4	5	17/5	B_{12}

[a] From Provasoli (1969).

system. There have been indications, however, that vitamins can affect more than just the growth rate of algal cells. Evidence has indicated that vitamins may also affect the reproductive capacity of some algae. However, not much is known about this at present and its overall effect on water quality is probably small.

C. Cations and Algal Growth

It is known that an increase in cation concentration can affect algal productivity, and hence eutrophication. Although there is little information available on the exact effect of changes in cation concentrations, work by Beeton (1969) on the Great Lakes has shown a shift in the types of algae present as a function of sodium and potassium levels. It is difficult in most cases to separate the effect of macronutrients from that of increased cation levels, but it does appear that some effects take place. In most reported instances it is not predominantly a change in growth rate that occurs, but more a change in the species or even type of algae that appear. In most situations, blue-green algae are favored at higher cation concentrations, and from a water quality perspective the blue-green algae are generally the most troublesome. It is well known that each use of water by humans increases the total ion concentration. Therefore, when used water is discharged back into receiving water systems, it usually has a higher ionic strength. If this affects algal growth, as suggested, it is possible that water systems are being influenced in the favor of blue-green algae rather than the less troublesome diatoms and green algae.

In conclusion, we have seen that metals, organic growth factors, and cations can affect algal productivity. It should be emphasized, however, that this is apt to occur only in isolated situations, mostly in oligotrophic or nonproductive systems. As more information is gathered, we may learn more about the effects of nutrients that are required only in low concentrations, but at present they are considered of little importance with regard to water quality.

VII. PLANT GROWTH AND EUTROPHICATION

We have now seen how nutrient cycles interact and how they are affected by human activity. Predictions can now be made as to the specific effects of various nutrient additions. First, however, we must describe the process of eutrophication and attempt to relate it to human activity.

A system that is tending toward a eutrophic state exhibits severe water quality problems. In general, with the creation of large amounts of algal biomass, the water becomes very turbid and aesthetically unpleasant. If the algal form is shifted to blue-green algae, then various toxins, tastes, and odors can be formed in the water, causing treatment problems if the water is to be used for drinking. A lake that has become eutrophic has a wide fluctuation in oxygen level. This is owing to algal growth that occurs during the daylight hours and creates supersaturated levels of oxygen in the lake. At night the oxygen level is readily diminished near zero by respiration. The fauna associated with such a lake system becomes severely stressed owing to these wild fluctuations in the concentration of oxygen. In addition, the heavy plant growth that occurs in the upper layers eventually dies and settles to the sediments, which then demand oxygen for stabilization. This oxygen demand can become severe, and in northern temperate lakes that freeze in the winter the oxygen can be completely depleted in the lake under the ice. As plant cells die and fall to the bottom, not only do they require oxygen for stabilization, but plant material tends to build up and create a pool of mineralizable nutrients.

The ecology of a water system undergoing eutrophication evolves, as does the trophic level. Plant forms that exhibit high growth rates usually dominate a eutrophic system, and often it is the blue-green algae or other nuisance types of algae that develop. Fish associated with a eutrophic system are generally forms that can tolerate large fluctuations in oxygen and heavy concentrations of algae. In general, it is observed that the ecosystem associated with a eutrophic body of water has a low diversity and is therefore unstable.

Algal growth can now be quantitatively defined as it relates to eutrophication. To do this we shall rely on physical and chemical equilibria as well as biological-growth kinetics. It should first be emphasized, however, that there is a basic difference between the terms "biomass" and "productivity." Biomass indicates the concentration of the algal standing crop in a system at any point in time and can be measured as the number of cells per unit volume, the chlorophyll content per unit volume, or the total suspended solids. Productivity, on the other hand, indicates the rate of biomass formation, and there is no clear connection between biomass and productivity. This is because the relationship is dependent on both space and time, so the reader should bear in mind that when we discuss productivity, we are talking about the rate of algal growth and consequently the rate of nutrient uptake. Biomass will be discussed in relation to the total algal standing crop that can be produced in a given system or is present at any given time in a given system.

A. The Stoichiometry of Algal Growth

Photosynthesis, as with any chemical process, can be modeled by relatively simple stoichiometry. Such modeling allows us to predict the amount of biomass that can be synthesized from the available nutrient concentrations. Odum (1971) proposed the following stoichiometry for photosynthesis:

$$106CO_2 + 16NO_3^- + HPO_4^- + 122H_2O + 18H^+ + \text{trace elements and energy}$$

$$\underset{\downarrow}{\overset{\uparrow}{}}$$

$$C_{106}H_{263}O_{110}N_{16}P_1 + 138O_2. \tag{17}$$

When this stoichiometry operates in the forward direction, that is, CO_2, NO_3, and PO_4 being converted to algal protoplasm and oxygen, it is called photosynthesis. When the process runs in the reverse direction, that is, algal protoplasm being converted back to the inorganic nutrients, it is called respiration. Equation (17) is a simplification of algal protoplasm, which can be variable depending on such factors as the growth media.

Equation (17) shows that the formation of oxygen during photosynthesis is in a direct molar ratio with the inorganic nutrients provided. As an example, the ratio of oxygen to phosphorus is about 138, and the ratio of oxygen to nitrate is about 9. Thus, if the nutrient content of a discharge to a system is known, the biomass potential of the system can be estimated, thereby allowing the engineer to evaluate the environmental consequences of the discharge. Similarly, if we have an indication of the algal biomass present in a water system, the possible respiration demand for oxygen during the dark hours can also be calculated.

The ratio of photosynthesis to respiration can also be used as an indicator of how much a system tends toward eutrophy. A balanced system, that is, a natural system, has balanced rates of photosynthesis and respiration, or the amount of oxygen produced by photosynthesis is very nearly equivalent to the amount consumed by respiration. Therefore, only very small fluctuations in the total dissolved oxygen content of such systems are observed. In a lake or a stream that has been enriched with nutrients, we find that the rate of photosynthesis is greater than the rate of respiration, owing to the development of a large algal biomass. Alternatively, if there are large organic discharges to a water system, such as domestic sewage, then respiration will tend to be greater than photosynthesis and a loss of dissolved oxygen will occur.

Some natural systems exhibit high productivity rates and yet still have a photosynthesis–respiration ratio close to one. A common example of this is a coral reef. In any case, a system may appear to be extremely productive, but a photosynthesis–respiration ratio of one indicates that it is not perturbed.

The productivity of a system is dependent on the input of inorganic nutrients, and the macronutrients are the most influential. It seems only logical that an approach to understanding the problem of eutrophication would be to evaluate the nutrient inputs to water systems. First, a decision must be made as to which nutrient limits algal production. We refer now to Liebig's law of the minimum, which states that crop yield (e.g., algal growth) is usually limited by one nutrient. This assumes, of course, that other nutrients are present in excess. We make this prediction by first considering the stoichiometry in Eq. (17), from which the ratios of nitrogen, phosphorus, and other compounds required for algal growth can be readily calculated. Noting, for instance, that carbon, nitrogen, and phosphorus are taken up in ratios of approximately 106 to 16 to 1, respectively, a water system could be analyzed to determine the ratio of the nturients present as a first approach in determining which nutrient limits algal production.

B. Determination of Trophic Level

Many attempts have been made to predict the trophic level of water systems as a function of either nutrient loading or some other measurable quantity. The original work by Sawyer (1947) has been noted, and his critical value of 10 mg m^{-3} has been used for years as a guide for predicting the consequences of nutrient enrichment in freshwater lakes. Vollenweider (1968) determined that the trophic level of lakes can be estimated based on a nutrient-loading-relationship that predicts the resultant concentration of phosphorus. From an evaluation of many lakes around the world, he determined that trophic classification limits should be set at 10 mg m^{-3} and 20 mg m^{-3} of phosphorus. Lakes exhibiting concentrations below 10 mg m^{-3} should be considered oligotrophic, and those containing greater than 20 mg m^{-3} should be designated as eutrophic. Vollenweider's model works relatively well on most freshwater lakes.

Dillon (1975) modified Vollenweider's model to account for flushing time in lakes. He proposed to calculate predicted phosphate concentrations as a function of loading rate, retention coefficient, hydraulic flushing rate, and mean depth. The following equation shows this relationship:

$$[P] = L(1 - R)/ZP, \qquad (18)$$

where L is the phosphorus loading rate in grams of phosphorus per square meter per year, R is the dimensionless retention coefficient, P is the flushing rate in meters of depth per year, and Z is the depth in meters. Relationship (18) allows us to predict the resultant phosphorus concentration in a lake.

The reader is referred to the original work for a discussion of the parameters used in the model. Once a prediction of the phosphorus level is made, its value can be compared with Vollenweider's criteria and a prediction made as to whether the lake will be eutrophic, mesotrophic, or oligotrophic. Much less work has been done using nitrogen loading to predict entrophication. This is because in most freshwater lakes nitrogen compounds are in excess, and thus phosphorus is limiting. In addition, the environmental movement of nitrogen is complex and more difficult to model.

REFERENCES

Alexander, M. (1961). "Introduction to Soil Microbiology." Wiley, New York.
Beeton, A. M. (1969). Changes in the environment and biota of the Great Lakes. *In* "Eutrophication: Causes, Consequences, Correctives," pp. 150–187. Nat. Acad. Sci., Washington, D. C.
Dillon, P. J. (1975). The phosphorus budget of Cameron Lake, Ontario: the importance of flushing rate to the degree of eutrophy of lakes. *Limnol. and Oceanogr.* **20**(1).
Eppley, R. W., *et al.* (1969). Half saturation constants for uptake of nitrate and ammonium by marine phytoplankton. *Limnol. and Oceanogr.* **14**, 912.
Goldman, C. R. (1964). Primary productivity and micro-nutrient limiting factors in some North American and New Zealand Lakes. *Vesh. Internat. Verein Limnol.* **15**, 365.
Leckie, J. D., and Stumm, W. (1970). Phosphate exchange with sediments; its role in the productivity of surface waters. *In* "Chemical Treatment" (E. Gloyna and J. Ecken, eds.). Univ. of Texas Press, Austin.
Odum, E. P. (1971). "Fundamentals of Ecology." Saunders, Philadelphia, Pennsylvania.
Provasoli, L. (1969). Algal nutrition and eutrophication. *In* "Eutrophication: Causes, Consequences, Correctives," pp. 574–593. Nat. Acad. Sci., Washington, D. C.
Sawyer, C. N. (1947). Fertilization of Lakes by Agricultural and Urban Drainage. *New England Water Works Assoc.* **61**, 109–127.
Snoeyinck, V., and Jenkins, D. (1981). "Water Chemistry." Wiley, New York.
Stumm, W., and Morgan, J. (1981). "Aquatic Chemistry," 2nd ed. Wiley (Interscience), New York.
Vollenweider, R. A. (1968). "Water Management Research." OECD Paris, Report No. DAS/CSI/68.27.
Weber, C. A. (1907). Aufbau and Vegetation der Moore Norddeutschlands. *Beibl. Botan. Jahrbuchern* **90**, 19–34.

Chapter 6

Microorganisms and
Water Quality

I. INTRODUCTION

This chapter will describe the various microorganisms involved in processes that affect water quality. We shall attempt to show how these organisms can modify the environment to either produce toxic products or purify polluted systems. To understand the role of microorganisms and their activity in aqueous systems, certain definitions should first be clarified.

The general classification of all microorganisms is the Protista, which is divided into the higher protists, eucaryotes, and the lower protists, procaryotes. The difference between the groups lies in the sophistication of the metabolic processes occurring within the cell. The eucaryotes include algae, fungi, and protozoa, whereas the procaryotes include bacteria and blue-green algae. Viruses are not included in these two classifications. Eucaryotic cells are generally $10-100\ \mu m$ in length, whereas procaryotic cells are usually about $1\ \mu m$ in length. Viruses, on the other hand, are seldom larger than $0.01\ \mu m$.

Microorganisms are generally ubiquitous in nature and are influential in regulating the overall quality of most water systems. They can be found occupying nitches in even the most stressed of environmental situations. Microorganisms have been isolated from waters containing very high salt concentrations, and they can grow in waters that have temperatures as low as 4°C and as high as 70°C. Microorganisms also exist in environments with varying amounts of oxygen.

Although each of the various environments discussed has an associated microflora, we are primarily interested in the microbial population of natural systems. All microorganisms grow and reproduce by consuming nutrients and building cellular material. The difference in many of the microorganisms that we shall consider lies in the fact that each can consume different forms of a particular nutrient. As an example, certain microorganisms use inorganic carbon as their carbon source for cellular construction, and certain others require specific forms of nitrogen for their metabolism. Therefore, depending on the relative proportion of certain types of microorganisms in a water system, we shall see a fluctuation in the various forms of these nutrients. This reflects on the overall quality of the entire water system.

II. EUCARYOTES

The eucaryotic cell has a more complex structure than does the procaryotic cell. Although this is not of particular interest from a water quality perspective, the difference should be understood. The predominant difference between the procaryotic and eucaryotic cell is the occurrence or absence of a defined nucleus.

A. Algae

Algae are a group of eucaryotes that are capable of photosynthesis. Photosynthesis was discussed in detail in Chapter 5 in relation to eutrophication in natural systems. Algae vary in size from microscopic plants as small as 10 μm up to seaweeds that can be 50 m long.

All algae require the inorganic nutrients carbon, nitrogen, and phosphorus and an adequate supply of sunlight. As discussed in Chapter 5, it is the metabolism of algae that often regulates the concentrations of nitrogen and phosphorus in natural systems.

Algae are differentiated primarily according to pigmentation, that is, green, brown, red, yellow-green, etc. These classifications come from the different pigments and/or different ratios of pigments found in various groups of algae. We have shown that excessive algal growth can degrade water quality, but it is also important to understand that some types of algae can be more troublesome than others. Green algae generally pose few problems in water systems, whereas other forms of algae can generate toxins. The blue-green algae, which we shall classify as procaryotic, are well known for their effect on water quality in this regard.

B. Protozoa

The second major group of eucaryotes is the protozoa. Protozoa are nonphotosynthetic, are extremely motile, and play important roles in degrading organic matter in water systems. They are also the principal predators upon certain bacteria, as we shall see later in this chapter. Protozoa are classified principally by their means of locomotion. Very little quantitative work has been done on protozoa populations in natural systems other than as possible biological indicators of aqueous toxins.

C. Fungi

The last major group of eucaroytes is the fungi. Fungi are nonphotosynthetic microorganisms that form large, branched threads. They are a complex group and are responsible for the degradation of many refractory materials in the environment. As an example, the degradation of cellulose is mostly facilitated by fungi.

Fungi are classified on the basis of sexual differentiation. Many well-known organisms such as mushrooms, yeast, and molds are fungi. Once again, from a water quality perspective, very little quantitative information is available on the role of fungi in environmental processes.

III. PROCARYOTES

Procaryotes are simple cells that contain no nuclear membrane. This group is dominated by the bacteria but also includes the blue-green algae. It is not clear taxonomically whether blue-green algae are photosynthetic bacteria or actually true algae. For the purposes of this book, however, we shall consider them to be blue-green algae, not blue-green bacteria.

A. Blue-Green Algae

Blue-green algae are known to cause water problems owing to the release of dissolved organic residuals and are generally considered to be objectionable when they occur in large concentrations. Physiologically, blue-green algae are extremely tolerant in environmentally stressed situations. They are resistant to high and low temperatures and are also capable, as noted in Chapter 5, of utilizing N_2 as a nitrogen source. This allows them to grow when other algal forms are starved for nitrogen. For a system in which concentrations of nitrate and ammonia are low, the blue-green algae are favored owing to their nitrogen-fixing capability.

B. Bacteria

Perhaps the most important group of protists with respect to water quality are the bacteria. These microorganisms vary in size from approximately 0.5 to 10 μm. Bacteria are capable of synthesizing cellular material from either inorganic or organic materials. Certain bacteria are known to function in the absence of oxygen (anaerobic growth), whereas others are strictly aerobic, requiring high levels of oxygen for growth. Many bacteria accelerate oxidation or reduction processes by consuming the energy released in the reactions. This catalyzes reactions that would proceed only at a slow rate in a sterile environment.

Although many forms of bacteria catalyze hydrogeochemical cycles, the most important group of bacteria with respect to water quality are those associated with the transmission of disease. The persistence of these microorganisms and their fate in the environment has been studied in great detail, because it relates directly to public health.

C. Viruses

The smallest microorganism to be considered here is the virus. Viral particles are essentially pieces of genetic material (nucleic acid), which can be either RNA or DNA. Viruses are not free-living units and must use a host cell to replicate. They penetrate the host cell and change the existing genetic material such that the cellular functions are converted to the replication of viral particals. The cell produces more viral particles and then bursts, liberating the viral particles to the environment for further attacks. Viruses attack many types of cells, including bacteria, algae, and animal cells. The fate of viruses in the natural environment is not well understood, owing to the difficulty in monitoring their occurrence. In Section V.C we shall examine the available information relating to the viability of viral particles in water systems.

We have now described the various types of microorganisms of interest in water quality. It is important to understand the metabolism of these organisms, for they affect almost all of the processes in natural systems. Also, once these processes are understood, they can then be engineered more effectively, for example, in waste treatment. We shall now consider the occurrence of various microorganisms in natural systems and observe the true diversity of their various metabolic processes, which can then be related to the resultant water quality.

IV. THE ECOLOGY OF MICROORGANISMS

The ecology of the microflora of natural systems is important, because it affects the ability of natural water systems to purify themselves. There are several texts available that describe the microbial ecology of natural systems in great detail [see Carpenter (1977) and Mitchell (1974)]; therefore, we shall focus on one aspect of the microflora, i.e., bacteria.

A. The Distribution of Bacteria in Natural Systems

Bacteria are found in water, air, and soil. The number of bacterial cells present in any particular location, as well as the particular types of cells present, gives an indication as to the ability of a system to purify itself. In addition, the level of environmental stress in an aqueous system can often be determined by observing the types and numbers of bacteria present.

Although it is beyond the scope of this book to evaluate all forms of bacteria, it is important to realize that certain general groups are common in most natural systems. The following list describes the forms of bacteria commonly found in natural systems (Mitchell, 1974).

(1)   *Pseudomonas.* Short, gram-negative, nonspore-forming, aerobic, motile rods that can produce fluorescent pigments. Most species oxidize glucose to produce acids.

(2)   *Rhizobium.* Gram-negative, aerobic, motile, nonspore-forming rods that form nodules on legumes. They can reduce nitrates to nitrites.

(3)   *Achromobacter.* Gram-negative, motile, nonspore-forming rods that cannot form pigments.

(4)   *Flavobacterium.* Short, gram-negative, motile rods that produce yellow, red, or orange pigments. They can decompose proteins.

(5)   *Micrococcus.* Spherical, gram-positive or sometimes gram-negative cells that occur in irregular groups. All species produce the enzyme catalase.

(6)   *Sarcina.* Spherical cells in regular packets, usually gram-positive, that produce white, yellow, orange, or red pigments.

(7)   *Bacillus.* Aerobic or facultatively anaerobic, motile, spore-forming rods that decompose proteins to yield ammonia.

(8)   *Clostridium.* Anaerobic, spore-forming rods.

Classification of bacteria is generally done on the basis of physiological differences, because morphological differences in bacteria are slight. Bacterial groups are classified by their ability to absorb certain dyes, utilize certain substrates, or produce certain products during synthesis.

B. Bacteria in Aqueous Systems

The flora of natural aqueous systems is relatively constant despite fluctuations in salinity and temperature. There is some difference in highly stressed systems, however, because only a few species can adapt. Willis *et al.* (1975) investigated the bacterial populations of three saline aquifers in Florida and North Carolina, and their data are shown in Table 6.1. It can be seen that the types of bacteria present essentially remained the same in spite of differences in location, temperature, and well depth.

Cherry, Guthrie, and Harvey (1974) evaluated bacterial populations in streams and ponds that received chemical and thermal waste. They showed that the most common isolates were *Bacillus, Flavobacterium, Brevibacterium, Achromobacter, Chromobacterium, Serratia, Sarcina, Escherichia, Streptococcus,* and *Pseudomonas.* The total bacterial count only varied from approximately 5×10^3 to about 9×10^4 cells ml^{-1}. Although there were some population differences between streams receiving chemical pollutants and a pond receiving thermal pollution, the total fluctuation was quite small. The total number of bacteria increased when the thermal

TABLE 6.1   COMPARISON OF THE BACTERIAL POPULATIONS OF THREE SALINE AQUIFERS[a]

PARAMETER	NORTH MONITOR, PENSACOLA, FLORIDA	OBSERVATION WELL NUMBER 7, WILMINGTON, NORTH CAROLINA	CALABASH SYSTEM, CALABASH, NORTH CAROLINA
Well depth (ft.)	1523	1050	1052
Temperature (°C)	34.8	22.7	—
pH	7.3	7.3	—
Number of facultative and aerobic bacteria per liter	—	60	141
Genera isolated	*Enterobacter* *Aeromonas* *Azomonas* *Bacillus* *Brevibacterium* *Micrococcus* *Staphylococcus*	*Acinetobacter* *Brevibacterium* *Corynebacterium* *Flavobacterium* *Desulfovibrio*	*Alcaligenes* *Arthrobacter* *Brevibacterium* *Flavobacterium* *Thiobacillus* *Desulfovibrio*

[a] From Willis *et al.* (1975).

effluent raised the temperature significantly.   Also, for certain cases in which chemical pollutants were increased to high levels in the stream, certain species of bacteria disappeared.   Guthrie, Cherry, and Singleton (1975) showed in a study of aquatic bacterial populations that high nitrate and phosphate levels do not have much effect on the stability of aquatic bacterial populations.   This further emphasizes the fact that unless the natural system is highly perturbed, the bacterial flora remains relatively constant and stable; in most cases the types of bacteria found are the same regardless of geographical location.

C. Bacteria in Sediments

The bacterial activity and resultant water quality of most aquatic systems is influenced in part by the sediments that underlie the system.   Although bacteria are found in soil systems as well as aqueous systems, their distribution in sediments is somewhat different.   Often the processes that occur in sediments that are mediated by bacteria are different than the corresponding processes that occur in the overlying water.   This is due to the slightly different type of environment found in sediments, for example, a greater surface area for reactions as well as a different oxygen regime.   In most cases we can consider aqueous systems to be homogeneous and completely mixed with respect to bacterial activity. Sediments, however, show a marked gradation from the water–sediment interface to the lower depths. This is owing in part to a strong oxygen gradient. Dale (1974) showed a strong relationship between bacterial populations and sediment properties in intertidal sediments. His studies showed that the population ranged from 10^8 to almost 10^{10} cells per gram of sediment and was highly correlated with grain size as well as nitrogen and carbon content. He found a substantial bacterial biomass as deep as 10 cm in the sediment. His calculated value for total biomass, that is, the total cellular material present, was in the same range as the larger macrofauna commonly found in the sediments.

The type of bacteria found in sediments depends to a large extent on the type of substrate. Bacterial activity using organic substrates is intimately related to the oxygen balance of the system as well as the loss of organic material. In addition, inorganic transformations can be performed by bacterial activity in sediments, for example, nitrification. Recall that nitrification refers to the oxidation of ammonia to nitrite and subsequently to nitrate. This process is mediated by two groups of bacteria: *Nitrosomonas* and *Nitrobacter*. *Nitrosomonas* mediates the oxidation of ammonia to nitrite, and *Nitrobacter* mediates the oxidation of nitrite to nitrate. These two groups of bacteria are classified as autotrophs, and they require a system that is low in organic carbon to operate because the carbon source for these bacteria is inorganic CO_2. For sediment systems in which the organic content is low and there is a sufficient oxygen supply, the oxidation of ammonia to nitrate proceeds rapidly.

Ardakani, Rehboch, and McLaren (1973) investigated the occurrence of nitrite and nitrate in soil columns. Figure 6.1 shows the loss of nitrite and corresponding increase in nitrate with depth in the sediments, reflecting the nitrification process in a soil sediment core.

Although nitrification is predominantly mediated by two select groups of bacteria, the opposite process can be mediated by a general group of bacteria. For soil systems in which the dissolved oxygen content is low, nitrate can be reduced to either ammonia or nitrogen gas by this process.

We have shown that microbial processes in sediments are often as important as those in the overlying water with respect to water quality. Both of these systems must be understood when dealing with water quality problems. Chemical transformations that do not occur in overlying waters proceed rapidly in the sediments owing to the different environmental constraints. In many cases, not only is the microbial activity different between overlying waters and sediments but the types of flora present also differ. This was discussed in Chapter 3 with regard to the transformation of various metals. In Section V.B bacteria that are used as indicators of water

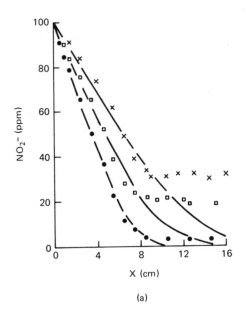

(a)

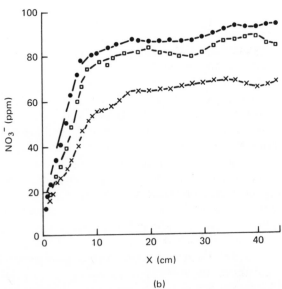

(b)

FIGURE 6.1   (a) Decline in the concentration of nitrite with depth in the soil column after 2 months for three flow rates in the soil: •, $F = 7.5$ cm h^{-1}; □, $f = 10.5$ cm h^{-1}; ✕, $f = 13.0$ cm h^{-1}.   (b) Nitrate distribution in the solution phase obtained by analysis of the samples used in (a). (From Ardakani, Rehboch, and Melaren, 1973.)

contamination will be shown to have different numbers and activities in the sediment regime as opposed to the overlying aqueous regime.

D. Relationships between Bacteria and Other Microorganisms

Another water quality phenomenon of interest is the interaction of bacteria with other microorganisms. This sophisticated interaction, whether symbiotic or antagonistic, is responsible for many of the subtle changes in water quality that occur in natural systems. Although many of these interactions are common in the natural environment, most of them are not well quantified with respect to resultant changes in water quality. Only those processes that deal directly with water quality will be considered here.

Perhaps one of the most important interactions involves bacteria and algae. Bland and Brock (1973) and Berland, Bolin, and Maestrini (1970) attempted to isolate and define those forms of bacteria associated with algae. Many bacterial forms were found to be commonly associated with various algal groups, including members of the genera *Vibrio, Flavobacter, Escherichia,* and *Pseudomonas.* Also commonly occurring were members of the genera *Sarcina, Staphylococcus,* and *Archromobacter.* This large group of heterotrophic bacteria are closely associated with the algae from which they were isolated. Much work has been done to delineate the actual interactions of these bacteria with the host plants, but the roles of all these microorganisms are not clear. The flora associated with the larger algae varies through the year depending on water temperature, but in general these groups are always intimately associated with the algae.

Some studies have shown that bacteria produce exudates that are required by algae for normal growth. Work by Provasoli and Pinter (1964), for example, shows that excretions from certain bacteria produce growth factors that are required for the morphology of two types of seaweeds, that is, when these seaweeds were cultured without bacteria they did not develop into their natural form. Waite and Mitchell (1976) isolated bacteria from a common seaweed and performed growth rate studies with and without the bacterial isolates. They showed that the effect on the productivity of the seaweed depended on the bacterial form that was reintroduced to the culture. Certain bacterial isolates improved the growth rate of the plant, whereas others inhibited the growth rate. This indicates that certain bacteria associated with these large algae can be symbiotic as well as synergistic. This work also indicates that when considering algal growth in aqueous systems, bacterial inputs must be also considered with respect to the synthesis of growth factors. Often preoccupation with inorganic nutrients as the sole regulator of algal photosynthesis produces misleading results.

The other aspect of bacterial interactions with algae, of course, is the degradation of plant material by bacteria. Chapter 5 described the energetics of the eutrophication process, which dealt primarily with excess plant production. Plant material must also be degraded in aquatic systems, and the ability of this material to be degraded is a function of the kinds of microflora that are present. Gunnison and Alexander (1975) investigated the susceptibility of 14 species of algae to microbial decomposition. They showed that although some algae were destroyed in short periods of time, others withstood microbial digestion for more than 4 weeks. They concluded that the cell wall of the plant was the main deterrent to bacterial degradation. This is not surprising considering that the cell wall of most plants is cellulose, which has a molecular structure that is difficult for bacteria to degrade.

The kinetics of algal degradation by bacteria is important with regard to resultant water quality. However, little quantitative data are available that allow for the prediction of degradation rates. Waite and Kurucz (1977) determined the bacterial degradation rates of an aquatic weed. They showed that the degradation followed the first-order model $y = me^{-kt}$, where y is the total plant material in the system at time t, m the initial plant material at $t = 0$, and k the first-order rate constant. They showed that the rate constant varied between 0.02 and 0.08 day^{-1} as a function of the nitrate concentration in the system.

We have now introduced the major microbial processes that occur in natural water and sediment systems. Organic matter degradation and inorganic transformations are important processes that are mediated by microbial populations in aqueous–sediment systems. The enumeration of bacterial population size as well as activity has been discussed with regard to its importance to changes in water quality. It should be emphasized that these natural populations are always present in aquatic systems and change only slightly depending on the environmental constraints of the system.

V. MICROORGANISMS AND PUBLIC HEALTH

We shall now discuss the most important aspect of microbial contamination in natural waters, that is, effects on public health. Bacteria, viruses, and protozoa are all known to cause disease; therefore, water quality must be maintained such that these diseases are not transmitted.

Microorganisms that cause disease in humans are termed pathogenic, and they can be introduced to water systems directly by sewage, animal excretions, or surface runoff. All of these mechanisms have been implicated in outbreaks of diseases around the world. It should be remembered that water and wastewater treatment processes were originally implemented for the prevention of disease.

A. Microorganisms and Disease

Before we discuss the more common diseases of public health concern, it is important to note that other mechanisms are often responsible for the transmission of disease. A good example would be the microorganism *Salmonella,* which can cause acute gastroenteritus and diarrhea. Table 6.2 shows the number of isolates of *Salmonella* found in various sources over a 4-yr period in the United States. Note that *Salmonella* represented only 2% of the total number of isolates from water supplies, whereas for isolation from other sources such as fowl they represented 46% of the total. Data have shown that *Salmonella* is often associated with several types of foodstuffs, including eggs and milk. Therefore, vectors other than water may be

TABLE 6.2 NUMBER AND PERCENT OF TOTAL *SALMONELLA* ISOLATES FROM NONHUMAN SOURCES IN THE UNITED STATES FROM 1963 to 1967[a]

SOURCE	NUMBER OF SALMONELLA ISOLATES	PERCENTAGE OF TOTAL ISOLATES
Fowl	16,172	46.43
Swine	2,791	8.01
Cattle	2,306	6.62
Other animals	2,474	7.10
Eggs	2,117	6.08
Dried milk	846	2.43
Other foods	1,406	4.04
Animal feed	3,703	10.63
Water	746	2.14
Miscellaneous	1,660	4.77
Unknown	609	1.75
Total	34,830	100.00

[a]From Cooper (1974).

responsible for the introduction of pathogenic microorganisms into human populations. These data also indicate that in fact only a small percentage of the total intake of *Salmonella* occurs from drinking water. Other microorganisms, however, are more readily transported by aqueous systems.

Table 6.3 shows the most commonly encountered diseases of public health significance along with the symptoms and mode of transmission of the disease. This table shows only those diseases caused by bacteria for cases in which the disease was transmitted by polluted water. In some instances the infection was due to contact, in others it was due to direct water intake. These diseases have different levels of severity, ranging from minor upset stomachs to acute symptoms. Many of these diseases are endemic in certain populations, especially among the lower economic classes and in the poorer countries. It should be noted that *E. coli* is listed as a disease-causing microorganism in Table 6.4. We shall see later that *E. coli* is also used as an indicator microorganism, and that it is considered to be nonpathogenic to humans at low concentrations. However, certain strains of *E. coli* have been observed to be pathogenic, but this only occurs when a large concen-

TABLE 6.3　BACTERIAL DISEASES OF PUBLIC HEALTH SIGNIFICANCE

BACTERIAL AGENT	DISEASE	SYMPTOMS	MODE OF TRANSPO
Salmonella	Salmonellosis	Acute gastroententis, diarrhea, nausea, and cramps	Sewage, polluted wate stormwater runoff, contaminated food
Shigella	Shigellosis	Acute diarrhea, high temperature, vomiting	Polluted water, conta nated food, person-person contact
Leptospira	Leptospirosis	Acute infection of kidney, liver, or central nervous system	Water contact, domes animals
E. coli	Enteropathogenic *E. coli*	Gastroenteritis, nausea, prostration, urinary infection	Polluted drinking wat stormwater runoff
Francisella tularensis	Tularemia	Fever, swollen lymph nodes, prostration	Contact with wood ticks, polluted drinking water
Vibrio comma	Cholera	Acute intestinal disease, rapid dehydration, usually results in death	Polluted drinking wat person-to-person contact, contamina food
Mycobacterium	Tuberculosis	Acute pulmonary disease	Polluted water, indus trial wastes, person person contact

tration of *E. coli* is present, such as occurs in hospitals or other institutions where sick individuals are confined.

In addition to bacterial diseases, viruses are known to cause various diseases of public health significance. Table 6.4 lists the common viruses that have been isolated, identified, and implicated as the causal agents of various diseases.

B. Indicators of Water Quality

From the preceding discussion it is apparent that many diseases can be transmitted by microbiological agents in water systems. It is important, therefore, to observe whether disease agents are present in a water system and at what level. It is then possible to evaluate the risk of the disease being transmitted to the general population.

To identify the presence of causal agents of disease, an in-depth microbial analysis must be performed. Assuming that reliable procedures are available for identifying each bacterial agent, the number of samples that would have to be collected on a continual basis would be tremendous. Thus, there develops an obvious need for an indicator species, that is, one organism that would be indicative of the presence of pathogens in the water. This indicator organism should have several properties, including that it be easy to identify and isolate from water systems. The other basic property is that

TABLE 6.4   VIRUSES FOUND IN ANIMAL FECES AND WASTEWATER[a]

BIOLOGICAL GROUP	ANTIGENIC TYPE	ASSOCIATED DISEASES OR SYNDROMES
Enteroviruses		
Polioviruses	3	Poliomyelitis, aseptic meningitis
Coxsackie virus group A	24	Hepangina, aseptic meningitis, exanthem
Coxsackie virus group B	6	Aseptic meningitis, myocarditis, pericarditis
Echoviruses	33	Aseptic meningitis, exanthem, gastroenteritis
Adenoviruses	31	Upper respiratory illness, pharyngitis, conjunctivitis
Reoviruses	3	Upper respiratory illness, diarrhea, exanthem
Infectious hepatitis virus(es)	1	Infectious hepatitis

[a] From Cooper (1974).

it be associated with human sources, so that its occurrence in a water system would indicate contamination by humans.

Escherich (1885) observed that *Bacillus coli* is present in high numbers in human excrement. He proposed that this organism might be used as an indicator of fecal contamination, for even in 1885 it was understood that human feces were responsible for many of the diseases common in those days. This finding was not used as a bacteriological standard for water until 1914, when the Public Health Service issued drinking water standards that regulated interstate water transport. The original standards were based on the *Bacillus coli* group and defined them as those microorganisms that were able to grow on lactose–peptone broth incubated for 2 days at 37°C and produce gas. Later work showed that the group of microorganisms isolated by this type of analysis inlcuded many forms that were ubiquitous in nature. There was an apparent need for being more specific with regard to isolating bacteria of human origin. A modification of the Eijkman test was utilized for this purpose, and it consisted of incubating samples in glucose at 44°C and observing gas production. In general, the nonfecal component was unable to produce gas at this elevated temperature.

These methods of determining the presence of coliform bacteria are essentially the same analyses that are used today. "Standard Methods for Water and Wastewater Analysis" now identifies the coliform group as all aerobic and facultative anaerobic, gram-negative, nonspore-forming, rod-shaped bacteria that ferment lactose with gas formation within 48 h at 35°C (Water Pollution Control Federation, 1975). This analysis, as noted previously, assays for total coliforms, which are those organisms in the family Enterobacteriaceae. This includes the genera *Escherichia, Klebsiella, Salmonella, Shigella, Aerobacter, Serratia,* and *Proteus.* Many of these genera are associated with the intestines of warm-blooded animals, in particular *Eschericia coli.* Members of all the other genera can be found in the environment as well as in warm-blooded animals. Work by Evison and James (1973) showed that the types of intestinal bacteria found in geographically distant areas are about the same. They compared the distribution of intestinal bacteria in both British and East African water sources, and with the exception of certain forms of *Citrobacter* and *Klebsiella* the coliforms were similar in both locations. It appears, therefore, that the use of coliforms as an indication of human contamination should be valid irrespective of geographical location.

1. Fecal Coliforms

Ever since the inception of the coliform standard, microbiologists have argued over the reliability of using the coliform group to measure the degree of human contamination. The most important question is whether the presence of coliforms indicates the presence of other pathogens in the system. In like manner the absence of coliforms should indicate a safe water system. Dutka (1973) suggested four criteria that indicator organisms should meet:

(1) to occur in greater numbers than the intestinal pathogens concerned;
(2) not to proliferate to greater extent in the aqueous environment than the enteric pathogens;
(3) to be more resistant to disinfectants and natural processes than the pathogens;
(4) to yield characteristic and simple reactions enabling as much as possible the unambiguous identification of the indicator.

It is possible to isolate pathogens from water systems in which coliform counts are negligible. Thus, there has been concern over the reliability of the use of coliforms as an indicator of the occurrence of pathogens. Dutka (1973), however, presents data on the relationship between coliforms and *Salmonellae* in the St. Lawrence River. He shows that in all the samples investigated there was a very high ratio of total coliforms to *Salmonellae*. It is obvious in Table 6.5 that the ratio of total coliforms to *Salmonellae* was so high in many instances that a large factor of safety exists. The table also indicates that the ratio of fecal coliforms to *Salmonellae* is also large and is perhaps a more reasonable indicator.

Warm-blooded animals appear to discharge large numbers of *E. coli* in their feces. Work by Geldreich (1967) showed that large numbers of warm-blooded animals do indeed produce fecal coliforms. Table 6.6 shows that a high percentage of individuals among humans, livestock, poultry, cats, and dogs contain fecal coliforms. It is postulated, however, that cold-blooded animals do not have a permanent intestinal flora, and that fecal coliforms are only present in cold-blooded animals owing to the high level in the surrounding water. Table 6.6 also shows that a high percentage of plants and insects were found to be contaminated with fecal coliforms. Fecal coliforms appear to be almost ubiquitous in nature, and they can certainly be introduced into water systems by animals other than humans; therefore, the use of fecal coliforms as an indicator of the presence of human feces may be invalid. Certainly, the occurrence of fecal coliforms in aqueous systems does not *prove* that the system is contaminated by domestic sewage.

2. Fecal Streptococci

The realization that even the presence of fecal coliforms cannot definitely identify human contamination of water systems has led researchers to look

TABLE 6.5 OCCURRENCE OF *SALMONELLA* AND INDICATOR-ORGANISM DENSITY IN WATER SAMPLES FROM THE ST. LAWRENCE RIVER[a]

COLIFORM MF DENSITY PER 100 ml	RATIO OF COLIFORMS TO *SALMONELLAE*	FECAL COLIFORM DENSITY PER 100 ml	RATIO OF FECAL COLIFORMS TO *SALMONELLAE*	FECAL STREPTOCOCCUS DENSITY PER 100 ml	RATIO OF FECAL STREPTOCOCCI TO *SALMONELLAE*
350	66,000	85	16,000	8	1500
220	42,000	4	760	2	380
150	7000	8	760	4	380
26	4900	9	1700	1	190
7	650	1	95	<1	95

[a] From Dutka (1973).

TABLE 6.6 PERCENTAGE OF SAMPLES
CONTAMINATED WITH FECAL COLIFORMS
FOR STUDIES OF HUMANS, ANIMALS, AND FISH[a]

FECAL SOURCE	PERCENTAGE OF SAMPLES CONTAMINATED
Warm-blooded animals	
Human	96.4
Livestock	98.7
Poultry	93.0
Cats, dogs, and rodents	95.3
Fish and their environment	
Tank fish (bluegills and carp)	1.1
Tank water (500-gal. capacity)	3.0
Little Miami River	
Water	31.0
Eight species of fish	30.8

[a] From Geldreich (1967).

for other bacterial indicators. Another principle component of human microflora is fecal streptococci, a group that includes all streptococci found in the intestines of warm-blooded animals. As with the coliforms, some of the streptococci are also ubiquitous in nature, including members of the groups *Enterococci, S. bovus, S. equinus,* and *S. fecalis.* The group *Enterococci* is the most common streptococci that is associated with humans.

The observation that makes fecal streptococci important as an indicator organism is that there appear to be unique ratios of fecal coliforms to fecal streptococci (FC–FS) for different warm-blooded animals. Although all warm-blooded animals contain both fecal coliforms and fecal streptococci, the relative size of the bacteria is different in humans than in other warm-blooded animals. The human population generally has a much higher ratio of fecal coliforms to fecal streptococci than do all other warm-blooded animals. Geldreich and Kenner (1969) noted that there was a difference in the FC–FS ratio of humans and other warm-blooded animals, and some of their findings are presented in Table 6.7. Note that humans tend to have high FC–FS ratios, usually in excess of 4.0, whereas all other warm-blooded animals tested had FC–FS ratios less than 0.7. The range of values between 0.7 and 4 is apparently not well defined with respect to the source of bacterial contamination. It should also be noted that the largest percentage of the fecal streptococci observed belonged to the *Enterococci* group.

This information suggests the possibility that FC–FS ratios can be used to

TABLE 6.7 BACTERIAL DENSITIES AND FECAL COLIFORM AND FECAL STREPTOCOCCUS DISTRIBUTIONS IN WARM-BLOODED ANIMAL FECES[a]

| FECAL SOURCE | NUMBER OF SAMPLES | DENSITY (gm^{-1})[b] | | FC–FS RATIO | TOTAL STRAINS EXAMINED | OCCURRENCE (%) | | | |
		FECAL COLIFORMS	FECAL STREPTOCOCCI			ENTERO-COCCI	S. BOVIS AND S. EQUINUS	ATYPICAL S. FAECALIS	S. FAECALIS LIQUI-FACIENS
Human	43	13,000,000	3,000,000	4.4	1067	73.8	None	None	26.2
Animal pets									
Cats	19	7,900,000	27,000,000	0.3	268	89.9	1.5	2.2	6.3
Dogs	24	23,000,000	980,000,000	0.02	585	44.1	32.0	14.4	9.6
Rodents	24	160,000	4,600,000	0.04	539	47.3	17.1	0.4	35.3
Livestock									
Cows	11	230,000	1,300,000	0.2	438	29.7	66.2	None	4.1
Pigs	11	3,300,000	84,000,000	0.04	296	78.7	18.9	None	2.4
Sheep	10	16,000,000	38,000,000	0.4	321	38.9	42.1	None	19.0
Poultry									
Ducks	8	33,000,000	54,000,000	0.6	328	51.2	48.8	None	None
Chickens	10	1,300,000	3,400,000	0.4	275	77.1	1.1	None	21.8
Turkeys	10	290,000	2,800,000	0.1	317	76.7	1.6	None	21.8

[a] From Geldreich and Kenner (1969).
[b] Median values.

indicate whether the bacterial contamination present in a water system is of human origin. Table 6.8 shows typical data from analyses made on wastewater and stormwater runoff. The FC–FS ratio was examined, and it can be seen that for all of the domestic wastewater samples examined, the FC–FS ratio was very large and was always greater than 4.0. The stormwater analysis showed FC–FS ratios to be much lower than 0.7, indicating the contamination was of nonhuman origin. This experiment indicates that, even though the coliform density was high in the stormwater runoff, indicating a possible pathogenic threat, the threat could be assumed to be minimal because the coliforms were of nonhuman origin.

Often the concentration of microorganisms in the water column and the sediments is different. It should be expected that pathogen concentrations would also be higher in the sediments than in the overlying water. Therefore, analyses of the water column to detect these pathogens or indicator organisms might not accurately indicate their presence in the sediments. Shellfish or other bottom-dwelling animals that live in the sediments would therefore be exposed to higher levels of pathogens (with the potential for concentration and transfer to humans) than would those animals that live in the water. Van Donsel and Geldreich (1971) used FC–FS ratios to evaluate the concentration of *Salmonella* in sediment samples. Table 6.9 shows that whereas all three cases considered tested positive for the presence of *Salmonella,* there were fewer positives for the samples in which the FC–FS ratios indicated nonhuman contamination. Although the data are somewhat sketchy, the trend indicates that when sewage contamination is present, there is a good possibility that *Salmonella* is present in the sediments.

3. Other Indicators

Recent improvements in bacteriological techniques have allowed more accurate monitoring of certain pathogenic microorganisms. Cherry *et al.* (1972) found *Salmonella* to be fairly common in "unpolluted systems." They proposed that *Salmonella* may indeed be free-living and multiplying in the natural environment. These determinations could be made only because the techniques for the isolation of *Salmonella* have improved. It is now possible to monitor *Salmonella* directly, without having to use indicator organisms to make estimates. Most other pathogens, however, are still difficult to isolate and their occurrence must be inferred from the presence of various indicator organisms.

Dutka, Chan, and Coburn (1974) proposed using the occurrence of compounds such as coprostanol and cholesterol, which are fecal sterols, to indicate human contamination of water systems. Humans and other

TABLE 6.8 DISTRIBUTION OF FECAL COLIFORMS AND FECAL STREPTOCOCCI IN DOMESTIC WASTEWATER AND STORMWATER RUNOFF[a]

WATER SOURCE	DENSITY (100 ml^{-1})[b]		FC–FS RATIO	TOTAL STRAINS EXAMINED	OCCURRENCE (%)			
	FECAL COLIFORMS	FECAL STREPTOCOCCI			ENTEROCOCCI	S. BOVIS AND S. EQUINUS	ATYPICAL S. FAECALIS	S. FAECALIS LIQUIFACIENS
Domestic wastewater								
Preston, Idaho	340,000	64,000	5.3	39	79.5	None	None	20.5
Fargo, North Dakota	1,300,000	290,000	4.5	50	100.0	None	None	None
Moorehead, Minnesota	1,600,000	330,000	4.9	50	90.0	10.0	None	None
Cincinnati, Ohio	10,900,000	2,470,000	4.4	428	71.5	2.8	1.6	24.1
Lawrence, Massachusetts	17,900,000	4,500,000	4.0	50	84.0	4.0	None	12.0
Monroe, Michigan	19,200,000	700,000	27.9	70	78.6	1.4	4.3	15.7
Denver, Colorado	49,000,000	2,900,000	16.9	70	85.7	11.4	2.9	None
Stormwater runoff								
Business district	13,000	51,000	0.26	1476	78.5	1.6	1.2	18.8
Residential	6,500	150,000	0.04	1158	80.0	0.5	1.4	18.1
Rural	2,700	58,000	0.05	445	87.4	0.5	0.2	11.9

[a] From Geldreich and Kenner (1969).
[b] Median values.

TABLE 6.9 DISTRIBUTION OF FECAL COLIFORM–FECAL STREPTOCOCCUS RATIOS IN THE OVERLYING WATERS FOR MUD SAMPLES POSITIVE AND NEGATIVE FOR THE PRESENCE OF SALMONELLA[a]

SAMPLES	FC–FS > 4.0			FC–FS 0.7–4.0			FC–FS < 0.7		
	NUMBER	%	FECAL COLIFORMS	NUMBER	%	FECAL COLIFORMS	NUMBER	%	FECAL COLIFORMS
Tested postive for salmonella (22)	10	45.4	280–610,000	8	36.4	41–24,000	4	18.2	6–212
Tested negative for salmonella (20)	3	15.0	100–2350	8	40.0	46–4100	9	45.0	4–13,000

[a] From Van Donsel and Geldreich (1971). © 1971 Pergamon Press, Ltd.

higher animals excrete high concentrations of cholesterol, which is a precursor to coprostanol, in their feces. The presence of either one of these sterols indicates potential contamination by humans. In addition, coprostanol is easily degraded by sewage treatment, and therefore if it is observed in a water system it is probably indicative of recent contamination. One of the current problems with this analysis is that there is no good data on the background levels of these sterols. Obviously, it is impossible to set safe standards at this point by using coprostanol. Also, to monitor coprostanol, elaborate extraction procedures in conjunction with chromatography analyses are required. This means that the analysis cannot easily be performed on a routine basis.

Kott et al. (1974) proposed the use of bacteriophages as indicators of human contamination in water systems. Bacteriophages are viral particles that attack bacterial cells and are usually specific for certain bacterial groups. Thus, if bacteriophages specific for E. coli are found in a water system, then E. coli must also be present. Bacteriophages are supposedly easy to monitor, and their presence should indicate not only the presence of E. coli but also of enteric viruses. In addition, bacteriophages do not appear to be as sensitive to chlorine as other organisms; an extra degree of safety should be added when using them as indicators.

C. The Fate of Intestinal Microorganisms in Natural Systems

Another requirement of indicator microorganisms is that they must behave in a manner similar to pathogens when discharged into the environment. In particular, the die-off rates must be comparable. Intestinal microorganisms usually die off rapidly when discharged to water systems. Many mechanisms have been postulated for the killing action, including sedimentation, solar radiation, predation, bacteriophage action, nutrient deficiency, and toxication.

Many researchers have evaluated the mechanisms that are responsible for the death of intestinal microorganisms, and it is not clear which process is dominant. Figure 6.2 shows the die-off rate of E. coli in seawater and the effect of further additions of E. coli. Negligible die-off occurs in filtered seawater and autoclaved seawater, whereas die-off proceeds rapidly in seawater that has been previously enriched with E. coli to develop a population of predatory microflora. It is therefore assumed that microbial predation is the most important mechanism, because none of the other components in seawater appear to be effective as killing agents.

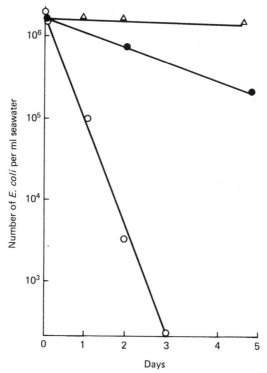

FIGURE 6.2   THE DIE-OFF RATE OF *E. COLI* IN FILTRATES OF SEAWATER.
The rate of kill increases with each addition of *E. coli* △, autoclaved seawater; ●,
filtered seawater; ○, filtered seawater followed by three additions of *E. coli*. (From
Mitchell and Morris, 1969. © 1969 Pergamon Press, Ltd.)

   Mitchell and Yankofsky (1969) showed that the microflora probably most
active as predators are the amoeba, especially *Vexillifera sp.*   Figure 6.3
shows the decline in concentration of *E. coli* in sterile seawater versus the
decline in seawater containing the amoeba *Vexillifera sp.*   Also shown is the
increased colony size of the amoeba with time.   This implies that *Vexillifera
sp.* is one of the principle predators upon intestinal bacteria and is chiefly
responsible for decreasing the concentration of intestinal bacteria in seawater.   This does not rule out other mechanisms such as solar radiation,
sedimentation, chemical stress, etc., but it does indicate that predation by
amoebas may be very important in the die-off of microorganisms in natural
systems.
   Reneau and Pehay (1975) investigated coliform movement through soils
caused by the percolation of septic tank effluents, and they showed that even
in marginal soils with low permeabilities, coliforms were not found at very

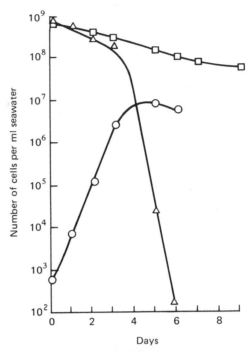

Days

FIGURE 6.3   LYSIS OF *E. COLI* CAUSED BY THE ADDITION OF A PURE CUL-
TURE OF *VEXILLIFERA* IN AUTOCLAVED SEAWATER; □, numbers of *E. coli* per
milliliter in sterile seawater inoculated with *E. coli;* △, number of *E. coli* per milliliter in
sterile seawater inoculated with *E. coli* and an amoeba lysin; ○, number of amoeba
per milliliter in sterile seawater inoculated with *E. coli* and an amoeba lysin.
(From Mitchell and Yankofsky, 1969.   Reprinted with permission from *Environ. Sci.
Technol.* **3,** 574–576. © 1969 American Chemical Society.)

great distances from the drainage field.   Their studies showed that after a
distance of 13 m the coliform concentration had essentially decreased to
zero.   Therefore, they postulated that bacterial movement does not proceed
far in ground waters.

Viruses should be considered in much the same manner as bacteria.
Viral particles also die off in the natural environment, but the mechanisms
involved are not as well understood as for the case of bacteria.   Akin *et al.*
(1976) evaluated the behavior of poliovirus I in marine water and found that
3 log units of virus infectivity were lost in 5 days at 24°C.   Figure 6.4 shows
that the concentration of poliovirus was attenuated quite rapidly in all of the
natural waters tested.   Filtered raw water, autoclaved water, and plain raw
water all exhibited approximately the same killing capacity, but the actual
inactivation agent was not determined.

Herrmann, Kostenbader, and Cliver (1974) studied the die-off rates of

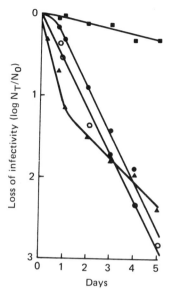

FIGURE 6.4   LOSS IN INFECTIVITY OF POLIOVIRUS I: ■, distilled water; □, filtered water; •, autoclaved water; ○, raw water. (From Akin *et al.*, 1976. © 1976 Pergamon Press, Ltd.)

two viruses—poliovirus I and coxsackie virus A9. Figure 6.5 shows the relative die-off rates of these viruses in sterilized lake water and natural lake water. The freshwater die-off rates seem to be slower than those in Fig. 6.4, that is, infectivity increased by 3 logarithmic units between the 8th and 18th days. In this particular study it was assumed that bacteria were the agent inactivating the viral particles by dissolving their protective protein coats.

Moore, Sagik, and Malina (1975) investigated the effect of suspended material on viral infectivity. They evaluated poliovirus and three coliphages in systems containing various amounts of suspended solids and found that no two viruses were absorbed in the same manner and that all had different infectivities. They concluded that suspended material can affect the infectivity of virus particles as well as the speed and efficiency of removal. Their work showed that virus particles removed by sedimentation can still be virulent, and this must be taken into consideration when determining water quality effects.

D. Quantification of Microbial Die-Off

Successful management of aqueous systems requires that mathematical models be made of all water quality processes. Although these models will

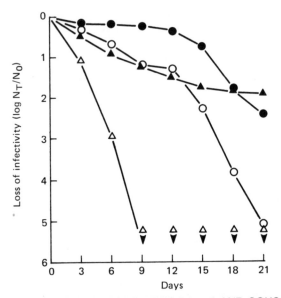

FIGURE 6.5   INACTIVATION OF POLIOVIRUS I (●, ○) AND COXSACKIEVIRUS
A9 (▲, △) IN LAKE WINGRA (OPEN SYMBOLS) AND IN STERILIZED LAKE WATER
(CLOSED SYMBOLS).   The arrow heads denote a value greater than that indicated
by the position of the triangle. (From Herrmann, Kostenbader, and Cliver, 1974.)

be treated in detail in Chapter 9, the quantification of the microbial die-off
process is integral to this chapter and will be treated here.   Models have been
created that reflect the die-off or loss of infectivity of pathogenic microorganisms in many water systems.   A few of the models will be described here so
that the reader can follow the general approach.

Pielou (1969) suggested a very simplistic model for representing the
natural die-off of microorganisms over time.   He proposed the simple
exponential function

$$N_t = N_0 e^{-\alpha t}, \tag{1}$$

where N_t is the number of microorganisms at any time t, N_0 the number of
microorganisms at time $t = 0$, and α the decay rate in units of t^{-1}.

Another model proposed by Frankel (1965) includes certain coefficients
so that the fitting to natural systems can be more easily performed.   The
equation suggested is

$$N_t = N_0(1 + nkt)^{-1/n}, \tag{2}$$

where k is the die-off coefficient and n the coefficient of nonuniformity.

Note that this form has more flexibility for fitting to environmental data by the manipulation of relevant coefficients.

Many simple functions of either a rational or exponential form can be postulated for modeling the decay of bacteria in natural systems. There are, however, few models available that attempt to represent the entire spectrum of processes that affect microbial decline in natural systems. Mitchell and Chamberlain (1974) proposed such a model. They created mathematical functions to represent each process known to affect microbial die off. These functions were then added together to form a partial differential equation modeling the change in microbial concentration with time and distance. The resulting equation is as follows:

$$U\frac{\partial C}{\partial x} - V_s\frac{\partial C}{\partial t} = \epsilon_y\frac{\partial^2 C}{\partial y^2} + \epsilon_z\frac{\partial^2 C}{\partial z^2} + \left(\frac{\mu_m s}{K_s + s} - a\right)C$$

$$-\left(\frac{\lambda_m P}{(K_c + C)Y_c}\right)C - (Kl(t)e^{-\alpha z})C - rC, \tag{3}$$

where C is the concentration of bacteria, U the current velocity, V_s the particle velocity, ϵ_y and ϵ_z the dispersion coefficients in the y and z planes, respectively, μ_m the maximum growth rate of the bacteria, K_s a half-saturation coefficient, s the concentration of the limiting nutrient, a the endogenous respiration rate, λ_m the maximum predation rate, K_c a half-saturation coefficient, P the concentration of microbial predators, Y_c the yield of predators on bacteria, K the die-off rate due to solar radiation, $l(t)$ the solar radiation flux, α the radiation attenuation coefficient, and r the die-off rate due to physical and chemical processes.

For the equation as written, an analytical solution would be difficult. However, Eq. (3) can be solved by making certain assumptions concerning the particular system under study, that is, certain terms can be omitted as insignificant under a given set of environmental constraints. In any case, Eq. (3) can be used to determine the relative effects of various parameters on microbial die off. [See Mitchell and Chamberlin (1974) for a discussion of the variables and coefficients.]

We have seen that indicator organisms are important in assessing the water quality of natural systems. The use of total coliform, fecal coliform, and fecal streptococcus indicators has proven useful in the last 40 yr for predicting contamination of water systems by humans. Although there is no doubt that the use of indicator organisms has many drawbacks, their usefulness has been proven over the years by the low incidence of water-borne disease in the United States. Apparently, coliform standards will continue to be used, and hopefully more representative models will be created so that better management tools will be available.

REFERENCES

Akin, E. W., Hill, W. F., Cline, G. B., and Benton, W. H. (1976). The loss of poliovirus I infectivity in marine waters. *Water Res.* **10**, 59–63.

Ardakani, M. S., Rehboch, J. T., and McLaren, A. D. (1973). Oxidation of nitrite to nitrate in a soil column. *Soil Sci. Soc. Am. Proc.* **37**(1), 53–56.

Berland, B. R., Bonin, D. J., and Maestrini, S. Y. (1970). Study of bacteria associated with marine algae in culture. III. Organic substrates supporting growth. *Mar. Biol.* **5**, 68–76.

Bland, J. A., and Brock, T. D. (1973). The marine bacterium *Leucothrix mucor* as an algal epiphyte. *Mar. Biol.* **23**, 283–292.

Carpenter, P. L. (1977). "Microbiology." W. B. Saunders, Philadelphia, Pennsylvania.

Cherry, D. S., Guthrie, R. K., and Harvey, R. S. (1974). Bacterial populations of aquatic systems receiving different types of stress. *Water Resour. Bull.* **10**(5), 1009–1016.

Cherry, W. B., Hanks, J. B., Thomason, B. M., Murlin, A. M., Biddle, J. W., and Croom, J. M. (1972). Salmonellae as an index of pollution of surface water. *Appl. Microbiol.* **24**(3), 334–340.

Cooper, R. C. (1974). Wastewater management and infectious disease. *J. Environ. Health* **37**, 217–224.

Dale, N. C. (1974). Bacteria in intertidal sediments: factors related to their distribution. *Linnol. and Oceanogr.* **19**(3), 509–518.

Dutka, B. J. (1973). Coliforms are an adequate index of water quality. *J. Environ. Health* **36**(1), 39–46.

Dutka, B. J., Chan, A. S. Y., and Coburn, J. (1974). Relationship between bacterial indicators of water pollution and fecal sterols. *Water Res.* **8**, 1047–1055.

Escherich, T. (1885). Die Darmbakterien des Neugeborenen und Säuglins. *Fortschr. Med. Virusforsch* **3**, 515–522.

Evison, L. M., and James, A. (1973). A comparison of the distribution of intestinal bacteria in British and East African water sources. *J. Appl. Bacteriol.* **36**, 109–118.

Frankel, R. J. (1965). Economic evaluation of water quality, an engineering–economic model for water quality management. SERL Rpt. No. 65-3, University of California, Berkeley, California.

Geldreich, E. E. (1967). Fecal coliform concepts in stream pollution. Paper presented at the Symposium on Microbial Parameters of Water Pollution, American Society for Microbiology, New York.

Geldreich, E. E., and Kenner, B. A. (1969). Concepts of fecal streptococci in stream pollution. *J. Water Pollut. Control Fed.* **41**(8), R336–R352.

Gunnison, D., and Alexander, M. (1975). Resistance and susceptibility of algae to decomposition by natural microbial communities. *Limnol. and Oceanogr.* **20**, 64–70.

Guthrie, R. K., Cherry, D. S., and Singleton, F. L. (1975). Effects of nitrate and phosphate concentration on natural aquatic bacterial populations. *Water Resour. Bull.* **11**(6), 1131–1136.

Herrmann, J. E., Kostenbader, K. D., and Cliver, P. O. (1974). Persistence of enteroviruses in lakewater. *Appl. Microbiol.* **28**(5), 895–896.

Kott, Y., Roze, N., Sperber, S., and Betzer, N. (1974). Bacteriophages as viral pollution indicators. *Water Res.* **8**, 165–171.

Mitchell, R. (1974). "Introduction to Environmental Microbiology." Prentice-Hall, Englewood Cliffs, New Jersey.

Mitchell, R., and Chamberlain, C. (1974). Factors influencing the survival of enteric microorganisms in the sea: an overview. *In* "Proceedings of the International Symposium on the Discharge of Sewage from Sea Outfalls," London.

Mitchell, R., and Morris, J. C. (1969). The fate of intestinal bacteria in the sea. *In* "Advances in Water Pollution Research," Proceedings of the Fourth International Conference, Prague, pp. 811–817.

Mitchell, R., and Yankofsky, S. (1969). Implication of a marine ameba in the decline of *Escherichia coli* in seawater. *Environ. Sci. Technol.* **3**, 574–576.

Moore, B. E., Sagik, B. P., and Malina, J. F., Jr. (1975). Viral association with suspended solids. *Water Res.* **9**, 197–203.

Pielou, E. C. (1969). "An Introduction to Mathematical Ecology." Wiley (Interscience), New York.

Provasoli, L., and Pinter, I. J. (1964). Symbiotic relationships between microorganisms and seaweeds. *Am. J. Bot.* **51**, 681.

Reneau, R. B., and Pehay, D. E. (1975). Movement of coliform bacteria from septic tank effluent through selected coastal plain soils of Virginia. *J. Environ. Qual.* **4**(1), 41–44.

Waite, T. D., and Kurucz, C. (1977). The kinetics of bacterial degradation of the aquatic weed. *Hydrilla sp. J. Air, Water, and Soil Pollut.* **7**, 33–43.

Waite, T. D., and Mitchell, R. (1976). Some benevolent and antagonistic relationships between *Ulva lactuca* and its microflora. *Aquat. Bot.* **2**, 13–22.

Water Pollution Control Federation (1975). "Standard Methods for Examination of Water and Wastewater," 4th ed. Water Pollution Control Federation, Washington, D.C.

Willis, C. J., Elkan, G. H., Harvath, E., and Dail, K. R. (1975). Bacterial flora of saline aquifers. *Ground Water* **13**(5), 406–409.

Van Donsel, D. J., and Geldreich, E. E. (1971). Relationships of Salmonellae to fecal coliforms in bottom seidments. *Water Res.* **5**, 1079–1087.

Chapter 7

Thermal Effects on Water Quality

I. INTRODUCTION

The term "thermal pollution" is used to describe water quality deterioration caused by inputs of heated water, mostly from industrial cooling processes. As the natural temperature regime of a water system changes because of thermal pollution, environmental systems become stressed. Although some members of aquatic ecosystems can adapt to the heated water, many are incapable of doing so and either die or are forced to relocate. However, the temperature range that can be tolerated by aquatic organisms is narrow. Therefore, when large volumes of heated effluent are introduced to a water system, at least some degree of stress quickly develops.

Many industries use water as a coolant. The principal user is the thermoelectric power industry, which consumes approximately 70% of all the water used for industrial cooling. Owing to the inefficiency of thermoelectric power generation, a large amount of energy (heat) is wasted, and this waste heat is absorbed by the cooling water. Each kilowatt hour of energy produced in a highly efficient coal-fired plant generates approximately 6000 Btu's, or two-thirds of the heat, as waste that must be removed by the cooling water. Nuclear power plants are even more wasteful (less efficient) because of the lower steam temperature in the throttle. Some nuclear facilities, in fact, waste 40% more heat than fossil fuel plants; the average nuclear plant transfers 10,000 Btu's of heat to the cooling water per kilowatt hour of electricity produced.

Because of the high standard of living in the United States, the demand for electric power is great. As the demand for electricity continues to increase, the number and size of thermoelectric power plants will also increase. It is estimated that in the near future the power industry will be using a large portion of the nation's total available freshwater runoff for cooling. Many areas of the country, especially the arid southwest regions, are now finding it difficult to supply the amounts of water required for cooling recently constructed thermoelectric power plants.

Large power plants are being constructed to achieve economies of scale. Generally, a 30 to 40% savings in production cost can be realized in power plants that can produce 1 million kW compared to those that produce 100,000 kW. The trend toward large power facilities, however, results in the generation of more concentrated loads of waste heat that must be handled by the environment.

II. SOURCES OF HEATED EFFLUENTS

Because the electric power industry is mainly responsible for the discharge of heated effluents, a brief review of the thermoelectric generation process is presented here. Most large thermoelectric power plants use Rankine steam – electric power conversion cycles, in which high-pressure steam is produced in boilers and then expanded through turbines to convert the thermal energy to mechanical energy. Figure 7.1 shows a schematic diagram of a typical power plant using a Rankine energy conversion cycle. If the diagram in Fig. 7.1 were a fossil fuel plant, then the following steps would take place in the generation of electric power.

(1) Water from the condenser is pressurized by pumping and heated in feedwater heaters.

(2) The pressurized water is heated in the boiler and converted to saturated steam.

(3) The superheated steam is expanded through the turbine, creating mechanical energy that drives the turbine and subsequently the generator.

(4) The resulting expanded, low-pressure steam is condensed in the

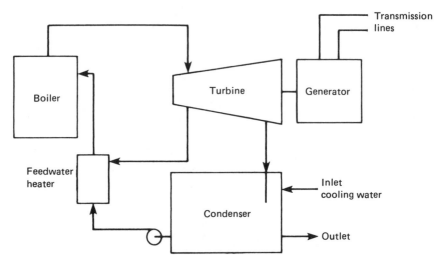

FIGURE 7.1   SCHEMATIC DIAGRAM OF A RANKINE POWER CONVERSION CYCLE.

condenser. The heat from this low-pressure steam must then be carried away from the power plant by the cooling water that circulates through the condenser, subsequently discharging high-temperature water.

There is a slight difference between the power conversion cycles associated with nuclear and fossil fuel power plants. The maximum operating temperature of a nuclear reactor is limited, as are the maximum pressures in nuclear pressure vessels. Therefore, nuclear steam boilers are generally not designed to produce superheated steam. This means that a smaller amount of energy is added to the power cycle, and hence heat rejection in the cycle is reduced. However, this smaller energy input is offset by the greater amount of work lost; therefore, nuclear power plants still discharge a large amount of heat energy per kilowatt hour of electrical energy generated. Thermal discharges from fossil fuel plants using superheated steam are generally 30 to 50% less than those from nuclear power plants with equivalent outputs of electricity. Some nuclear plants, however, use moisture separation and reheating cycles to improve efficiency, which partly compensates for the reduced efficiency in the basic Rankine cycle.

There are several ways of estimating the thermal discharge from a given power plant. All of the estimation methods take into account plant efficiency, cycle efficiency, boiler efficiency, etc., and all assume that the amount of thermal energy discharged is equal to the total energy input minus the thermal energy loss minus the electrical energy produced. The total energy input to the utility is equal to the heat value of the fuel times the quantity of the fuel, which is usually expressed as kilocalories or Btu's per kilowatt hour. For modern steam power plants, these values range from 9000 to 11,000 Btu kW^{-1} h^{-1}. These same calculations can be made for nuclear plants, except that nuclear plants do not have thermal losses through the stacks, which would normally be included in the lost energy term.

All electric power generators use condensers to cool low-pressure steam that has entered the condenser from the turbine. The steam entering the condenser then expands so that its concentration is below saturation, with approximately 1000 Btu lb^{-1} steam required for condensation. The quantity of water required for this condensation is inversely proportional to the temperature rise allowed in the condenser. For example, Fig. 7.2 shows that the quantity of cooling water required for an 800 MW fossil fuel plant operating at normal efficiency (35%) with an allowed temperature rise of 10°F is approximately 800,000 gal min^{-1}. This tremendous volume of water is required for only an 800 MW plant—current practice is to build much larger facilities. Thus, the potential for discharging large volumes of heated effluent can be seen.

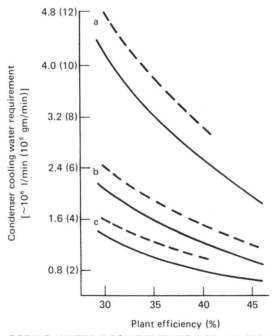

FIGURE 7.2   CODING WATER REQUIREMENTS FOR AN 800-MW ELECTRIC POWER PLANT FOR SELECTED ALLOWED TEMPERATURE RISES IN THE CONDENSER: a, 5.6°C; b, 11.1°C; c, 16.7°C; ——, fossil fuel; ––––, nuclear. (From Environmental Protection Agency, 1974).

In the past most electric power plants used once-through cooling systems. In this case cooling water is withdrawn from a supply source, circulated through the condensers, and then discharged back into the environment. When power plants were small and widely dispersed, no appreciable temperature increase in the cooling water was noted. As power plants become larger and more numerous, producing increased volumes of heated effluents, controls on the temperature of water discharges may have to be implemented. In principle, by limiting the temperature rise in the condenser, the discharge temperature of the rejected cooling stream can be controlled, thereby minimizing its environmental impact. Increasing the temperature in the condenser to minimize cooling water requirements is generally unacceptable, however, because it increases the temperature of the discharged water. This is contrasted with the low temperature rise in the condenser that results when using a larger flow of cooling water, which is also unacceptable because a large amount of biological material builds up in the cooling system. This seemingly unsolvable dilemma of using either large flows with small temperature increases or small flows with large temperatuare increases can only be solved by the use of auxiliary cooling systems. Therefore, most new electrical generating systems use some sort of auxiliary system, generally evaporative cooling. However, even with the use of evaporative cooling, large volumes of makeup water will be required to compensate for evaporation losses. As an example, evaporation losses from a 1000-MW nuclear power plant can be as much as 200,000 gal day^{-1}, and because this water has been evaporated, it is not available for downstream use. In addition, evaporation losses remove nearly pure water, thereby concentrating dissolved substances in the remaining water.

Many cooling systems are being designed as "closed cycle," that is, the water is recycled through the cooling system back to the condensers. The returned cooling water therefore has a higher temperature than that for a once-through cooling system. Because the temperature of the returned cooling water increases, so does the condensation temperature and pressure on the steam side of the condenser. In turn, the back pressure on the turbine tends to increase, resulting in an overall decrease in plant efficiency. Because of this loss of efficiency, more thermal energy must be supplied to produce the same amount of electrical output. The end result is increased fuel consumption and higher costs. The various types of auxiliary cooling systems utilized in the electric power industry will now be described.

1. Natural Draft Towers

These auxiliary cooling facilities are large chimneys that provide a draft that pulls air over the surface of the water to be cooled. They are generally hyperbolic in shape to achieve an optimum aerodynamic and structural design. Towers of this type generally operate efficiently, and because they have no moving parts they are relatively maintenance free. However, because of the height required for good air draft, natural draft towers are unusually large and aesthetic considerations have often precluded their use. For natural draft cooling towers currently in operation, typical decreases in cooling water temperature range from 24 to 38°F.

2. Mechanical Draft Towers

Mechanical draft towers are similar to natural draft towers except that the air flow is forced or induced by mechanical equipment. The structure itself does not have to be as large as for a natural draft cooling tower, so these towers are generally more aesthetically pleasing. They do, however, require energy to move air across the cooling water. In general, induced draft towers are more effective than natural draft towers. Figures 7.3 and 7.4 show typical natural and induced draft cooling towers.

3. Spray Ponds and Power Spray Systems

When sufficient land space is available, it is often advantageous to use a spray pond as an auxiliary cooling system. In this system cooling water is sprayed into the air through nozzles and collected in a pond. The ponds require very little maintenance and are efficient when sufficient area is

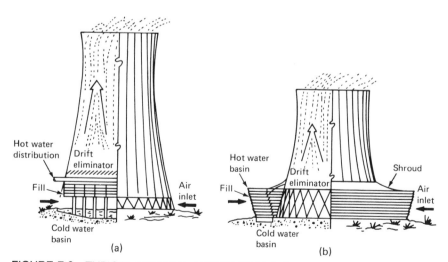

FIGURE 7.3   TYPICAL NATURAL DRAFT COOLING TOWERS: (a) counterflow; (b) crossflow.

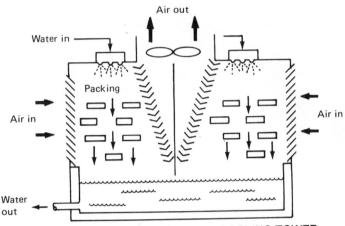

FIGURE 7.4   AN INDUCED DRAFT COOLING TOWER.

available for rapid cooling.   If spray nozzles are not utilized and cooling occurs only through conduction, convection, and radiation, then much larger surfaces of water are required.   In general, a cooling pond without spraying requires approximately 2 acre MW^{-1} for a fossil fuel plant and 3 acre MW^{-1} for a nuclear plant.

As mentioned previously, when any auxiliary cooling facility is used in an electric power facility, large volumes of water are lost by evaporation.   This requires a large volume of makeup water and tends to concentrate suspended and dissolved solids in the remaining water.   The concentrations of these substances in the cooling water are also increased by the addition of chemicals in the power plant to control fouling and corrosion.   Most auxiliary cooling systems therefore require periodic wastage of the concentrated cooling stream (blowdown), which often creates another water pollution problem.   Hence, many closed-cycle cooling streams require some treatment before they can be discharged to the environment.

III.   TEMPERATURE EFFECTS ON BIOLOGICAL SYSTEMS

Most plants and animals associated with water systems are incapable of regulating their internal body temperature.   Therefore, the temperature of such organisms fluctuates in accordance with ambient temperature of the environment.   This means that little buffering capability is available in aquatic ecosystems for withstanding large fluctuations in water temperature. If ambient water temperatures change owing to additions of thermal effluents, for instance, a shift in species diversity usually results, and organisms incapable of functioning optimally at the new temperature eventually disappear.

A. Natural Communities

The natural temperature regime of water systems is a function of latitude, altitude, and regional climatic patterns. This can be seen in Fig. 7.5, which shows average groundwater temperatures in the United States. Note that there is almost a 40°F change in ambient temperature from the northernmost to the southernmost point of the United States. Therefore, some temperature adaptability is required of various biological groups. There are certain organisms, of course, that occupy regimes of extreme temperature. Some microorganisms, for example, are known to be metabolically active in temperatures approaching freezing, whereas at the other extreme, for example, hot springs, bacteria have been isolated from 88°C water and algae from 73°C water (Brock, 1967). Although it is apparent that some forms of life can exist in these extreme conditions, one should consider these as unusual

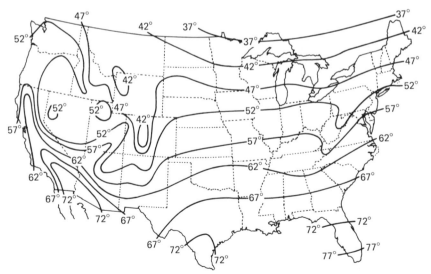

FIGURE 7.5   AVERAGE GROUNDWATER TEMPERATURES IN THE UNITED STATES (°F). (From Collins, 1925.)

groups of organisms. This is verified by the fact that only a few species are found in extreme-temperature environments.

The effects of temperature fluctuations are different for different organisms. Death can occur when temperature fluctuations become severe, and in the case of warm water enzymes begin to be deactivated at temperatures above 35°C. For higher life forms, the effects of temperature fluctuations can be more subtle. Slight changes in respiratory activity and motor coordination occur as temperature increases. In these cases change in the structure of the biological community is slow and can be observed only with careful study.

One of the more common physiological responses to temperature change is the alteration of osmotic regulation. Freshwater organisms must maintain ions inside their bodies at higher concentrations than in the surrounding water. This is done by concentrating salt within body membranes. As the temperature of the ambient water increases, this process is altered, affecting the ionic balances of the organisms.

When the temperature shifts in natural systems are slow and relatively small in magnitude, most of the biological community can adapt. The degree of adaptability to temperature changes, however, is different for different organisms and generally depends on their natural optimum temperature. This temperature tolerance for organisms has been designated as cold stenothermal, eurythermal, and warm stenothermal. The largest tolerance range is associated with those organisms classified as eurythermal. Organisms normally classified on either the warm or cold side of eurythermal have small thermal tolerance ranges, and therefore warm water discharges to cold environments, for example, would tend to be more harmful to these groups than to the eurythermal group.

There is a different and often more complex response to temperature changes for higher-order aquatic organisms. In this chapter we shall focus on the lower life forms, for example, bacteria, protozoa, fungi, and algae, rather than more complex organisms such as fish. As organisms become more complex, with more complex life cycles, different phases of their existence require different temperature regimes. Data on the lethal temperature for a full-grown fish, for example, does not imply that the juvenile or larval stage of the animal will survive that temperature range. This means that a wide range of temperatures can be tolerated by complex organisms depending on the stage of the life cycle. Also, higher organisms that are mobile (i.e., fish) are able to move to temperatures that are more suitable for their life processes. Evaluating their occurrence or distribution in a system as a function of water temperature is therefore difficult. As an example of the complexity of temperature requirements for fish, Table 7.1 shows the optimum temperature ranges for various freshwater fish.

TABLE 7.1   OPTIMUM TEMPERATURE RANGES FOR SELECTED
FRESHWATER FISH

SPECIES	ACCLIMATION TEMPERATURE (°C)	UPPER LIMITING TEMPERATURE (°C)
Cyprinus carpio (carp)	26	34
Esox lucius (northern pike), juvenile, 5.0 cm	25	32.25
minimum	27.5	32.75
	30	33.25[a]
Esox masquinongy (muskellunge), juvenile,	25	32.25
5.0 cm minimum	27.5	32.75
	30	33.25[a]
Fundulus chrysotus (golden topminnow), adult	35	38.5
Gambusia affinus affinus (mosquito fish), adult	25	37.0
	30	37.0
	35	37.9[a]
Gambusia affinus holbrooki (mosquito fish),	15	35.5
adult	20	37.0
	30	37.0
	35	37.0[a]
Ictalurus nebulosis (brown bullhead)	5	27.8
	10	29.0
	15	31.0
	20	32.5
	25	33.8
	30	34.8
	34	34.8
Ictalurus punctatus (channel catfish),	26	36.6
juveniles, 44–57 days old	30	37.8
	34	38.0
Ictalurus punctatus (channel catfish),	25	35.5
juveniles, 11.5 months old	35	38
Lepomis macrochirus (bluegill), adult	15	30.5
	20	32.0
	25	33.0
	30	33.8
Lepomis megalotis (long-ear sunfish),	25	35.6
juveniles, larger than 12 mm	30	36.8
	35	37.5

[a] Ultimate upper incipient lethal temperature.

B. Primary Productivity

The effects of natural temperature fluctuations on primary productivity are difficult to define. This is owing primarily to the fact that for those periods of the year when the temperature of the water is warm, the photoperiod is also longer. Therefore, the separation of effects on productivity is difficult. Studies of natural systems do indicate, however, that as the ambient temperature increases, the gross productivity of biological systems increases. It can therefore be anticipated that if a heated effluent increases the ambient water temperature by a small amount, an increase in productivity would result.

Temperatuare changes usually affect several parameters in natural biological communities. Experiments must therefore be planned so that only temperature is varied to ascertain absolute effects on primary productivity. Kevern and Ball (1965) studied biological communities established in replica streams in which only temperature was varied (from 20°C to 25.6°C). Dissolved oxygen measurements were taken to monitor primary productivity, and no significant difference in productivity was observed between the two ambient temperatures. Gross productivity was slightly greater at the higher temperature, but the net productivity values were similar because of increased respiratory metabolism at the higher temperature. On the other hand, Phinney and McIntire (1965) observed increases in both net and gross oxygen production in periphytic communities that were transferred to different temperature regimes and light intensities. Their work indicates that the effect of temperature on photosynthesis depends on the amount of illumination available.

Felfödy (1961) showed that the photosynthetic response of a natural algal community dominated by *Diatoma elongatum* was affected by temperature. The experiments, conducted at 5°C intervals from 10 to 30°C, showed that there was a broad optimum with respect to oxygen evolution extending from 10 to 20°C. At higher temperatures the photosynthetic response was markedly reduced, whereas at lower temperatures the species diversity of the community shifted in favor of diatoms.

Several studies have evaluated the behavior of aquatic macrophytes in response to temperature fluctuation. Anderson (1969) investigated the response of several macrophytes to thermal changes in Maryland. His study showed that the occurrence of the algae *Ruppia maritima* diminished owing to the addition of heated effluent, whereas the algae *Potomogeton* was more tolerant to elevated temperature and replaced the other algae as the dominant form.

Changes in plant activity owing to temperature shifts can be more subtle than just the observed loss of oxygen production. Studies by Waite and Mitchell (1976) showed that there is an interaction, both beneficial and antagonistic, between algae cells and the associated microflora. They showed that this interaction is probably integral to the overall metabolism and primary productivity of algae. Working with the associated microflora of the macrophyte *Ulva lactuca,* they showed that the plant's primary productivity in constant temperature environments tended to vary according to the outside ambient water temperature. It was postulated that temperature changes affected the associated microflora in some manner and subsequently affected the primary productivity of the plant.

Studies of the effects of temperature fluctuations on entire aquatic communities have also been made. Beyers (1962) investigated temperature effects on sediments containing algae, bacteria, protozoa, and invertebrates. He showed that the productivity response, as measured by total ecosystem respiration, was relatively unaffected by temperature changes from 14 to 32°C. In contrast to these observations, an ecologically unstable sewage community showed major effects owing to temperature fluctuations. From this work he concluded that the more stable an ecosystem, the less it is affected by temperature fluctuations. In general, stable ecosystems tend to resist environmental perturbations to a greater extent than unstable systems.

Many studies have shown that most algae have a limited temperature range in which they can grow and survive. In addition, there is an optimum range at which algae exhibit the highest primary production. In most situations the optimum primary production temperature ratio changes in response to ambient temperature fluctuations. There is, however, no indication that the upper limit of metabolism can be changed by environmental conditions. Also, it has been suggested that the tolerance of algae for high temperature is fixed, perhaps genetically, and cannot be altered by environmental changes. The optimum metabolic range for algal cells is generally located near the upper end of the tolerance range of the organism. In natural systems one usually finds that a species is only common or dominant when the temperature regime is near its particular optimum. Bott, Patrick, and Vannote (1973) classified algae into groups on the basis of preferred temperature regime. They classified algae as cold-water stenotherms (0 to 15°C), temperate stenotherms (15 to 25°C), and warm-water stenotherms (>25°C). Hustedt also included a classification for algae that can withstand large temperature fluctuations, and uses the terms meso-stenotherms for those forms that can withstand a temperature fluctuation of 10°C, meso-eurytherms for those forms that can withstand temperature fluctuations of 15°C, and eu-eurytherms for those that can withstand temperature changes of 20°C or more.

Temperature effects on algal populations, and hence primary productivity, are characterized by shifts to different species. In temperate water systems communities are composed of various types of algae, including green algae, blue-green algae, and diatoms. As ambient temperatures are increased to the upper end of the tolerance range, green algae often become dominant. As the temperature continues to increase, the green algae give way to the blue-green algae, which are the only forms found in warm or hot springs. Certain groups of blue-green algae have the highest temperature tolerances of any of the algae. This shift in species dominance caused by temperature increase is important, because many groups of algae, especially the blue-green algae, pose water quality problems. When blue-green algae grow in large numbers, high concentrations of organic residue are generated, causing taste and odor problems. Therefore, although warm water discharges may not significantly affect the overall primary production of water systems (in fact they may increase the primary production for small temperature differences), they can shift the species dominance of the algae to the unfavorable blue-green forms. If the thermal discharges are sufficient to elevate temperatures above those tolerated by blue-green algae, then no primary production will take place in the receiving waters.

The second component of natural food chains is the zooplankton, which can also be affected by temperature fluctuations. These effects include changes in metabolic respiration rates, reproduction rates, and feeding rates. As for algae, zooplankton have different optimum temperatures for each of these functions, but most of the temperature studies with zooplankton have been undertaken in natural systems. Zooplankton generally tend to be active in the summer, with major growth and reproduction occurring in the warmer months. When water temperatures fall below 10°C, most zooplankton become dormant and exhibit little activity. Therefore, in temperate regions it is not uncommon to find all zooplankton reproduction and growth compressed into a 200-day period of activity.

Insects also represent an important component in natural food chains, and they too are affected by temperature changes. In a study conducted by Bott, Patrick, and Vannote (1973) to evaluate a river with large temperature fluctuations in Ontario, the number of species of both caddis flies and mayflies increased at the warmer downstream stations. It was also noted that the occurrence of the stonefly species most tolerant of cold water decreased as the water became warmer. The temperature gradient in the test streams varied from 13 to approximately 20°C. It was shown that a continuous gradient of species gain or loss occurred as one moved from the colder headwaters to the warmer downstream waters. It would therefore be expected that a cold water system receiving thermal discharges would show a change in the species diversity of insects as well as other organisms. The magnitude of the temperature change caused by a thermal discharge would dictate the actual shift in species diversity. As an example, Vincent (1967) showed that a 4°C rise in mean annual temperature and a 6°C rise in maximum temperature resulted in a reduction of the average yearly diversity index of insects. He also found that, although diversity decreased during the period of warmer temperatures, the biomass, or standing crop, increased as the temperature increased. This finding is indicative of stressed ecosystems in which a particular type of organism acclimates to severe conditions and dominates. This is an unstable system, however, and therefore not favorable.

C. Aquatic Bacteria

Although much information is available on the effect of temperature changes on higher organisms, little information can be found on temperature changes altering bacterial activity in natural systems. Bacteria are ubiquitous in aquatic systems and usually possess broad tolerances to dissolved oxygen, pH, and temperature shifts. Bacteria that grow at temperatures greater than 55°C are termed thermophiles, and those that can grow at temperatures greater than 70°C are called extreme thermophiles. Some of these organisms were discovered in natural hot spring environments.

The microorganisms of most interest from a water quality perspective are enteric organisms that are associated with humans. In the rest of this chapter we will emphasize the behavior of human pathogens such as *Salmonella* and *Shigella* and indicator organisms such as *E. coli* and fecal streptococci.

Gallagher *et al.* (1965) showed that the die-off rate for *Salmonella* decreased as the temperature of a natural stream decreased. They showed that freshly introduced cultures of *Salmonella* could be recovered 22 miles downstream from the point of introduction in warm water, but when the temperature was colder *Salmonella* cultures were isolated as far as 73 miles downstream. Geldreich *et al.* (1968) observed 99% die-off for *Salmonella typhimurium* in stormwater stored at 10°C for 14 days. When the temperature was raised to 20°C, 99% die-off occurred in only 10 days. These studies indicate that enteric pathogens would persist in natural systems longer in cold water. Although it is not clear which bacterial inactivation mechanism predominates in natural systems, it appears that it is more effective at warmer temperatures. Therefore, extrapolation of these findings to the case of discharges of heated effluents concludes that enteric organisms should not survive for long periods of time at the warmer temperature.

The use of enteric organisms such as *E. coli* as indicators of pollution is based on the assumption that they do not grow in the natural environment. The presence of these organisms presumably indicates human fecal contamination. Hendricks (1972) recently studied the growth of some representative members of the enterobacteriaceae at various temperatures in sterilized river water. He showed that all organisms tested reached their maximum growth rate at 30°C. *E. coli* and *E. aerogenes* exhibited doubling times of 34 and 33 h, respectively, whereas the enteric pathogens examined had doubling times of 76 to 90 h. When the temperature was lowered to 20°C, the organisms still grew but with generation times approaching 1000 h. He was therefore able to show that indicator organisms do indeed grow in natural systems, but that their growth rate is slow in cold water environments. If, however, the ambient temperature is raised to between 25 and 30°C, the doubling times of enteric organisms increase to such an extent that they may not be useful indicators of recent human fecal contamination.

IV. BENEFICIAL USES OF THERMAL DISCHARGES

Theoretically, a beneficial use of waste heat (for example, effluent discharge after a condenser rise of 10° and 30°F) must either reduce a thermal water pollution problem or provide economic compensation that offsets the cost of cooling devices. Many projects have been undertaken that attempt to utilize one of these criteria but none have yet gained widespread acceptance. Problems arising from the transportation of the waste heat and the marketing of the product derived from waste heat have precluded widespread use.

There are several methods of directly using waste heat, including low-temperature heat and energy additions to secondary sewage, agriculture, and aquaculture. It should be noted, however, that in most cases temperatures in excess of 100°F (which is typical of effluents from electric power condenser systems) are required.

Recent studies have shown that biological sewage treatment can be made more efficient by raising the temperature with low-grade waste heat. In some cases this can increase treatment efficiency by a factor of 10. Waste heat from condenser systems is also being utilized in agricultural applications. It appears that hot wastewater can be used in open field irrigation as well as in temperature-controlled greenhouses and animal shelters. Using heated water for crop irrigation prolongs the growing season, increases production, and in general improves the quality of the crops. In recent experiments the use of heated effluent for subirrigating certain crops has also been evaluated. Results indicate that crop yield is significantly greater when heated effluents are used for irrigation.

However, problems have been encountered with the use of cooling water that contains chemicals used for corrosion, scaling, and biofouling control. In addition, radioactive contamination of agricultural products is possible when utilizing cooling water effluent from nuclear power plants.

As mentioned previously, waste heat can be used indirectly for increasing crop productivity by heating greenhouses. As an example, heating costs for a commercial greenhouse using fossil fuels can be as much as $2000 to $11,000 per acre depending on location. Therefore, if waste heat from a nearby cooling system were used, a large savings in heating cost would be realized. It is known that production in a greenhouse environment is far in excess of normal yields for the same crops grown conventionally. One of the major problems associated with the use of waste heat in greenhouse environments is the marketing of the produce itself. Also, it is estimated that greenhouses capable of using one-fourth of the waste heat from a 100-MW power plant would require a capital investment of about $25 million and occupy some 101 ha.

Perhaps the most widely studied application of waste heat has been direct use in aquaculture. Japanese researchers pioneered this area, and they have had test programs operative since the early 1960s. In fact, by 1967 aqua-cultured fish products in Japan represented 6% of the total catch and 15% of the total value.

Of the 2500 known species of fish, less than 1% have been cultured successfully. Studies have shown that yields of a few hundred kilograms per hectare can be sustained by using only the natural food elements available in water systems. If supplemental feeding is done, yields can be increased to as high as 1800 to 2700 kg ha^{-1} yr^{-1}. Catfish are the most commonly cultured species in the United States owing to their relative tolerance to the harsh environmental conditions that usually develop in high-density fish ponds.

Optimal feeding conditions are not the only important consideration for high production rates. Optimal temperatures for sustained growth can be realized by using waste heat. As an example, oyster farms on Long Island have been producing oysters on a commercial basis using the heated effluent from a power generation facility. Using this waste heat the normal 4-yr oyster growing cycle has been reduced to almost $2\frac{1}{2}$ yr, because the normal 4–6 month period of slow growth can be minimized.

Many other installations around the country have been investigating the use of thermal additions for developing high-yield aquaculture systems. Catfish, shrimp, lobsters, eel, yellowtail, sea bream, and whitefish have all been cultured. The major problems facing these facilities right now deal with the marketing of the produce that they generate. It appears that once the scientific problems with respect to generating large masses of fish are overcome the marketability of the system must then be determined. Certainly if the produce cannot be readily sold for a reasonable profit there will be less incentive for developing such aquaculture systems.

Aquaculture systems do not substantially reduce the temperature of thermal effluents. In addition, temperatures during warm summer months in many areas of the country can be so high as to preclude aquaculture. Therefore, the profitability of such endeavors might be so low that the utilities generating the heat effluent would not be interested in pursuing aquaculture systems. Also, the presence of any chemical or radioactive contaminants in the waste effluent would hinder aquaculture processes. In general, the economics of aquaculture applications have not yet been adequately demonstrated for large-scale commercial processes.

REFERENCES

Anderson, R. R. (1969). Temperature and rooted aquatic plants. *Chesapeake Sci.* **10,** 157.

Beyers, R. J. (1962). Relationship between temperature and the metabolism of experimental ecosystems. *Science* **136,** 980.

Brock, T. D. (1967). Microorganisms adapted to high temperatures. *Nature* **214,** 882.

Collins, W. D. (1925). "Temperature of Water Available for Industrial Use in the United States," U. S. G. S. Water Supply Paper 520-F. U. S. Geological Survey, Washington, D. C.

Environmental Protection Agency (1974). "Development Document for Effluent Limitations Guidelines and New Source Performance Standards for the Steam Elecrtric Power Generating Point Source Category," EPA Report No. 440|1–74|029-a. Environmental Protection Agency, Washington, D. C.

Felfödy, L. J. M. (1961). Effect of temperature on the photosynthesis of a natural diatom population. *Ann. Biol. Tihany* **28,** 95.

Gallagher, T. P., Thomas, N. A., Hogan, J. E., and Spino, D. F. (1965). "Report on Pollution of Interstate Waters of the Red River of the North" by the U. S. Department of Health, Education, and Welfare. Robert A. Taft Sanitary Engineering Center, Cincinnati, Ohio.

Geldreich, E. E., Best, L. C., Kenner, A., and Van Donsel, D. J. (1968). The bacteriological aspects of storm water pollution. *J. Water Pollut. Control Fed.* **40,** 1861.

Hendricks, C. W. (1972). Enteric bacterial growth rates in river waters. *Appl. Microbiol.* **24,** 168.

Bott, T. L., Patrick, R., and Vannote, R. L. (1973). The effects of natural temperature variation on riverine communities. *In* "Effects and Methods of Control of Thermal Discharges," part 3. Report to the Congress by the Environmental Protection Agency.

Kevern, N. R., and Ball, R. C. (1965). Primary productivity and energy relationships in artificial streams. *Limnol. Oceanogr.* **10,** 74.

Phinney, H. K., and McIntire, C. D. (1965). Effect of temperature on metabolism of periphyton communities developed in laboratory streams. *Limnol. Oceanogr. 10,* 341.

Vincent, E. R. (1967). A comparison of riffle insect populations in the Gibbar River above and below the Geyser Basins , Yellowstone National Park. *Limnol. Oceanogr.* **12,** 18.

Waite, T. D., and Mitchell, R. (1976). Some benevolent and antagonistic relationships between *Ulva lactuca* and its microflora. *Aquat. Bot.* **2,** 13–22.

Chapter 8

Potential Impact of Air Contaminants on Water Quality

by J. E. QUON†,*

I. INTRODUCTION

Both natural and anthropogenic sources of contaminants contribute to water pollution. The major man-made sources are well known, and although the contributions of natural sources are recognized, the extent of their contribution is not accurately known. The potential contributions of air contaminants to water quality have received only cursory attention, and a controversy over the extent and magnitude of this potential contribution still exists.

Apart from an intrinsic interest in environmental chemical cycles, the recent need for a methodology to assess the atmospheric contribution to water pollution is prompted by the need to develop regional water quality management plans under Section 208 of the Federal Water Pollution

† Deceased.
* Department of Civil Engineering, Northwestern University, Evanston, Illinois.

Control Act Amendments of 1972. Other types of planning and engineering also require that the various sources of water pollution be identified and quantified.

Several investigators have concluded that the effect of atmospheric contributions on water quality is small (Huff, 1976; Johnson, Rossano, and Sylvester, 1966). However, a number of studies† support the thesis that atmospheric contributions have a significant impact on water quality and that, in some instances, atmospheric contributions of certain substances to a body of water can be larger than the stream contributions of these substances.

The general linkages between air and water environments can be thought of in the following terms. First, the impact of atmospheric contaminants on water quality can be divided into several components. The major components are the flux, or rate of deposition, of a contaminant to a land surface, the fraction of the deposited material that ends up in a stream or body of water, and the rate at which the deposited material reaches the body of water. Thus, the atmosphere is a potentially large source of water pollutants, but its impact on water quality depends on the fraction and rate of deposition of the material that reaches the water. An estimate of these various components is needed to provide an overall evaluation of the relative importance of each air contaminant to the deterioration of water quality. In addition, one must also consider all the significant reactions and interactions of a specific air contaminant in and with the air, land, and water environments.

An air contaminant can be transferred directly or indirectly onto the surface of a body of water. In direct air–water transfer, exchanges at the air–water interface for gases, liquids, and solids are possible during both dry and wet weather periods and are referred to as dry or wet deposition. Dry deposition includes the fallout of atmospheric particulates owing to gravity and the turbulent deposition of both particulates and gases during dry weather periods. Wet deposition includes rainout, washout, snowout, and sweepout. Rainout and snowout refer to processes occurring within clouds, that is, particles forming condensation nuclei upon which water or ice, respectively, condenses. Washout and sweepout refer to the removal of material below the cloud level by falling rain or ice, respectively. (In the case of an ice cover over a body of water, the various exchanges are still possible, but the cover acts as a reservoir and a time delay is introduced in the contribution.)

---

† Bormann *et al.* (1968); Brezonik (1972); Fisher *et al.* (1968); Gatz (1975); Hasler (1970); McMahon, Denison, and Fleming (1976); Murphy (1974, 1976); Murphy and Doskey (1975); Winchester and Nifong (1971).

The indirect transfer of contaminants from air to water is modified by the transport, transformation, and storage of contaminant on land. Storage can introduce a substantial time delay between the time when a contaminant reaches the land and the time when the contaminant shows up in the water. For example, contamination of groundwater owing to the salting of highways can take up to 20 years to manifest itself, provided that there is not a direct connection between the runoff waters and the groundwater aquifer. The presence of snow can also introduce a time delay. In this case the contributions of precipitation to stream flow are delayed as well.

Transformations (biological, chemical, and physical) are ever present in the air, water, and land environments. When, for example, hydrogen sulfide is introduced into the air it generally ends up in the water environment in the form of sulfates. Once in the water, the sulfate ions can be converted back to hydrogen sulfide under anaerobic conditions and be released to the atmosphere. Ammonia in the air can reach the water as NH_4^+ or as the nitrite or nitrate ions. In the water the nitrogen species are incorporated into the flora and fauna and eventually released to the atmosphere in the forms of ammonia or nitrogen gas. These are examples of the nutrient cycles that were discussed in Chapter 5.

The storage of air contaminants deposited on land provides ample opportunities for the contaminants to be transformed into other chemical forms prior to their reaching waterways. Materials deposited on different vegetative surfaces or soils can also undergo chemical transformations, and the rates of deposition differ depending on the type of surface. In many instances the transformations and their rates are important to the amelioration or aggravation of water quality problems.

Generalized models of the transfer of material from air to water environments are still under development. As yet, however, the important modes of transfer have not been satisfactorily identified and quantified. We shall attempt to identify the potentially important modes of transfer of air contaminants to water environments, using a set of selected water quality indices, and then evaluate the relative magnitudes of the contributions.

The general approach to assessing the contributions of air contaminants to water quality will consist of developing a conceptual model to describe the various modes of transfer, followed by an evaluation of the relative magnitudes of the contributions and, hence, the relative importance of each mode of transfer. The development of the conceptual model depends on an initial choice of water quality indices and on the translation of expressions of air quality into expressions used as indicators of water quality. The modeling process must be comprehensive enough to include the set of water quality indices of interest. The water quality indices usually of greatest interest to regional water quality management and planning include total organic carbon, inorganic carbon, suspended solids, volatile suspended solids, *ortho*-phosphorus, organic phosphorus, ammonium nitrogen, nitrite plus nitrate nitrogen, and organic nitrogen.

This chapter will focus on the link between the air and land environments, that is, the flux of atmospheric substances to the land. Air contaminant concentrations and water quality indices vary with both time and space. Further, some of the potential linkages between the air and water environments have seasonal characteristics. This chapter, however, will consider the flux of atmospheric contaminants to the land on an annual basis. Estimates for atmospheric contributions are given in terms of annual deposition rates, for example, kilograms of nitrate nitrogen per hectare per year. The land-to-water link and water quality modeling will be discussed in Chapter 9, but a cyclic atmosphere – land – water environmental model is beyond the scope of this book.

In this chapter estimates of total dry deposition are referenced to air quality that is, the concentrations of the total suspended particulates and the various gases in the air. Estimates of total wet deposition are referenced to air quality and to the amounts and rates of precipitation. The chemical composition of suspended particulates and the concentrations of the various gases in the atmosphere are used to estimate the fluxes of the various substances to the land. The chemical composition of rainwater is used as a check on the estimated fluxes. The example evaluation presented will draw principally upon studies conducted in Illinois, but the methodology for estimating regional loads of air contaminants is generally applicable to any particular site if the required air quality data and chemical compositions of suspended particulates are available.

II. REVIEW OF PREVIOUS STUDIES

Potential atmospheric contributions to water pollution have been evaluated for various bodies of water. Johnson, Rossano, and Sylvester (1966) determined that dustfall contributed only minimally to the deterioration of the water quality of reservoirs in the Seattle, Washington area, even though total dustfall fluxes to the surrounding area ranged from about 600 to 3600 kg $ha^{-1} yr^{-1}$. McMahon, Denison, and Fleming (1976) demonstrated that the atmosphere may be a large source of nutrients for the Great Lakes. Regional loads of sulfate determined by field measurements, excluding dry deposition of gases, on Lake Superior and Lake Huron – Georgian Bay were found to be 10.9 kg S $ha^{-1} yr^{-1}$ and 15.8 kg S $ha^{-1} yr^{-1}$, respectively. It was also found that approximately one-fifth to one-third of the phosphorus input to Lake Michigan was from precipitation (Murphy, 1974, 1976; Murphy and Doskey, 1975). A study of nitrogen and phosphorus sources for Wisconsin surface waters (Hasler, 1970) estimated that the direct transfer of nitrogen and phosphorus from precipitation falling on water surfaces contributed approximately 8.4 kg N ha^{-1} yr^{-1} and 0.19 kg P ha^{-1} yr^{-1}, or 8.5 and 1.2% of the total nitrogen and total phosphorus inputs, respectively. Brezonik (1972) reported that precipitation onto lake surfaces in the United States contributes from 1.8 to 9.8 kg N ha^{-1} yr^{-1} and from 0.15 to 0.60 kg P ha^{-1} yr^{-1}. Atmospheric inputs of nitrogen and phosphorus may contribute significantly to the stimulation of plant growth and the acceleration of eutrophication as described in Chapter 5.

Winchester and Nifong (1971) estimated the potential atmospheric impact on the pollution of Lake Michigan by trace elements by using weight ratios of emissions of selected contaminants from large point sources in the Chicago and Northwestern Indiana regions. Based on an arbitrary transfer of aerosols from the air to the lake of 10%, the atmosphere could be a significant source of lake pollution for the trace elements copper, cobalt, manganese, nickel, vanadium, titanium, and zinc, and possibly also for iron, sulfur, chromium, and aluminum. Gata (1975), in a study of aerosol deposition into Lake Michigan, concluded that between 3 and 15% of atmospheric emissions of selected elements from large point sources in the Chicago and Northwestern Indiana areas enter the lake, because the plumes traverse the lake. Further, a comparison of estimated atmospheric inputs by total (wet and dry) deposition into the lake with tributary stream inputs into the lake showed that the estimated ratios of atmospheric inputs to tributary inputs for the elements iron, lead, and titanium were approximately 1.6, 2.0, and 1.2, respectively.

Several investigators have examined the potential impact of atmospheric contaminants on stream water quality in different watersheds. In a study of water quality in two basins downwind from a large urban industrial area (St. Louis), Huff (1976) inferred that the atmospheric contribution to stream water pollution was, in general, relatively small in comparison with the total annual stream loads of various anions and cations other than nitrate and zinc. The weight ratios of the total (wet and dry) deposition of atmospheric fallout on each basin to the total stream load for selected substances were evaluated. These ratios varied from 0.006 for Na$^+$ to 2.23 for Zn^{2+} and for most substances analyzed ranged from 0.02 to 0.08. The large ratio for zinc of 2.23 obtained for one basin and the relatively large ratios for NO$_3^-$ of 0.43 and 0.51 for two basins, however, implied that atmospheric deposition of zinc to one basin was more than twice the annual stream load in the basin's creek and approximately 50% of the stream loads for nitrates in both basins. Because not all of the material deposited in the basin reaches the stream, a small weight ratio of deposition to stream load would imply that the atmospheric contribution to water pollution is negligible, and only weight ratios greater than a few tenths indicate the potential for significant effects on water quality.

Fisher *et al.* (1968) and Bormann *et al.* (1968) examined the impact of atmospheric precipitation (into the drainage area) on stream water quality in a forested area of New Hampshire with respect to sulfate, nitrate, and ammonium. Wet deposition of these substances ranged from 10.0 to 14.2 kg S ha^{-1} yr^{-1} for SO_4^{2-}, from 1.6 to 2.2 kg N ha^{-1} yr^{-1} for NH_4^+, from 1.5 to 4.1 kg N ha^{-1} yr^{-1} for NO_3^-. For the watersheds studied, stream loads ranged from 9.8 to 18.1 kg S ha^{-1} yr^{-1} for SO_4^{2-}, from 0.2 to 0.9 kg N ha^{-1} yr^{-1} for NH_4^+, and from 1.1 to 1.7 kg N ha^{-1} yr^{-1} for NO_3^-. Thus, precipitation could account for most of the sulfate load carried by the streams, and the wet deposition of nitrate and ammonium exceeded the stream outflows of these substances.

In addition to wet deposition measurements in forested areas of the United States, other investigators have evaluated the air-to-land component and measured the flux of various substances from the air to the land or land cover. The flux of SO_2 and other sulfur compounds from the atmosphere to land has been widely studied. Chamberlain (1960) calculated that the deposition of SO_2 in rainfall over Britain is 11 kg S ha^{-1} yr^{-1}, using an atmospheric SO_2 concentration of 20 $\mu g/m^3$. Davies (1976) measured the wet deposition of SO_2 in Sheffield, England as 10 kg S ha^{-1} yr^{-1}. For agricultural areas in Britain, Fowler (1978) estimated dry deposition of SO_2 to be 36 kg S ha^{-1} yr^{-1}.

Mayer and Ulrich (1978) found that removal rates of atmospheric sulfur depend on the type of air–land interface. Measurements of total (wet and dry) depositions of atmospheric sulfur in central Europe were reported to be 23, 47–51, and 80–86 kg S ha^{-1} yr^{-1} for a bare soil, a beech forest, and a spruce forest, respectively.

Measurements of nitrate and ammonium due to wet deposition in Central Europe range from 1 to as high as 6–8 kg N ha^{-1} yr^{-1} for NO_3^- and from 1 to greater than 8 kg N ha^{-1} yr^{-1} for NH_4^+ (Oden, 1976). In the United States, wet depositions of ammonium and nitrate were 2.0–2.1 kg N ha^{-1} yr^{-1} and 2.8–4.5 kg N ha^{-1} yr^{-1}, respectively (Bormann *et al.*, 1968; Fisher *et al.*, 1968). Wagner and Holloway (1975) measured the total (wet and dry) deposition of several metals onto land in northwest Arkansas. Total deposition rates were 3.2 kg Na^+ ha^{-1} yr^{-1}, 3.2 kg K ha^{-1} yr^{-1}, 14.4 kg Ca^{2+} ha^{-1} yr^{-1}, and 1.3 kg Mg^{2+} ha^{-1} yr^{-1}. Dry and wet depositions of other selected trace elements from the atmosphere to the air–land interface are available (Likens *et al.,* 1967; Peirson *et al.,* 1973; Peyton *et al.,* 1976; Struempler, 1976).

The estimated fluxes of selected substances from the air to the land we have reported are large in many cases, indicating that the atmosphere is a potentially large source of water contaminants. However, as noted previously, not all of the material that is deposited on a region of land will reach a body of water. For example, the fraction of nitrates deposited by precipitation on the Kaskakia river basin in Illinois that reaches the stream was estimated by Harmeson, Sollo, and Larson (1971) to be approximately 50%. For a large point source of air contaminants whose plume traverses an expansive water surface, estimates of the fraction of material that impinges on the water surface range from 3 to 15% (Gatz, 1975) to 25% (Skibin, 1973) for the overall deposition of particles and from 12 to 25% for the dry deposition of particles (Sievering, 1976; Sievering and Williams, 1975). The impact of atmospheric contaminants on water quality is generally greater for clean rather than polluted bodies of water and for waters with large rather than small surface areas.

The large variations in the fluxes of various substances from the atmosphere and the resulting stream loads reported are partly caused by differences in geographic factors and/or the circumstances under which the measurements were made. But the controversy surrounding the impact of the atmospheric deposition of various substances on water quality may also be, in part, attributed to the poorly defined nature of the problem. Apparent contradictions may simply be references to different aspects of the problem, or to similar problems but under different circumstances.

III. BASIC DATA

In assessing the potential impact of air contaminants on water quality, the concentrations of air contaminants must be translated into contributions to the water quality indices of interest. Estimation of the fluxes of air contaminants to land that have a potential impact on water quality requires data on the concentrations of air contaminants and their chemical composition and data on rainfall and the chemical composition of rainwater. In this chapter estimates of the wet and dry deposition onto land of various substances are referenced to air quality, that is, the concentrations of total suspended particulates (TSPs) and the various gases in the air and the chemical composition of TSPs.

The potential contributions of the flux of phosphorus from the atmosphere to a body of water are principally owing to the phosphorus content of particulate matter. The dry deposition of phosphorus onto land can be estimated from the dustfall and phosphorus content of TSPs. The wet deposition of phosphorus due to the scavenging of TSPs by rain is small. Atmospheric contributions of the various forms of nitrogen are largely associated with the wet deposition of ammonium, nitrate or organic nitrogen constituents of the suspended particulate matter, and the gaseous ammonia and nitrogen oxides present in the atmosphere. A washout ratio (Englemann and Slinn, 1970; Inter-agency Committee, 1975), a dimensionless ratio of the concentration of a substance in rain to the concentration of the same substance in air, is used to estimate the fluxes to land of the various forms of nitrogen associated with the scavenging of TSPs. Solubility data, equilibria considerations, and washout coefficients (Slade, 1968) for ammonia and nitrogen oxides are used to develop estimates of wet deposition associated with the scavenging of these gases by rain. The chemical cycle for sulfur and the wet and dry deposition of sulfur dioxide have been widely studied. Some information on the wet and dry deposition of sulfur is considered and included in this analysis to serve as a check on the general validity of the methods of estimation used. The chemical composition of rainwater is also used as a check on the estimates of wet deposition for the various constitutents.

As noted previously, the water quality parameters of interest include total suspended solids, volatile suspended solids, total organic carbon, inorganic carbon, ortho-phosphorus, total phosphorus, ammonium nitrogen, nitrite plus nitrate nitrogen, organic nitrogen, and acidity. For these water quality indices, the principal air contaminants of interest are total suspended particulates, the volatile and water soluble fractions of TSPs and the organic carbon, organic phosphorus and *ortho*-phosphorus, ammonium nitrogen, nitrite plus nitrate nitrogen, and organic contents of TSPs. The composition of TSPs for selected constituents of interest is shown in Table 8.1. Rainwater concentrations of suspended solids, organic and inorganic carbon, organic phosphorus, ammonium nitrogen, nitrite plus nitrate nitrogen, and organic nitrogen are needed along with rainfall data to assess the wet deposition of air contaminants onto land. The concentrations of selected constituents of precipitation are shown in Table 8.2.

In addition to data on the chemical composition of TSPs and rainwater, the atmospheric concentrations of particulate matter and the various gases are required to evaluate the fluxes of the various materials to land. Typical concentrations of TSPs and the gases of interest for both urban and global background areas are given in Table 8.3. With the exception of particulates, carbon dioxide, and possibly ammonia, typical urban concentrations

TABLE 8.1   SELECTED CONSTITUENTS OF TOTAL SUSPENDED
PARTICULATE MATTER

CONSTITUENT	PERCENTAGE BY WEIGHT	REFERENCE
Benzene-soluble organics	5.1 – 6.6	McMullen, Faoro, and Morgan (1970)
	2.2 – 5.0	Shanty and Hemeon (1963)
	5.1 – 14.2	Williamson (1973)
Volatile fraction	25	Stern (1976)
Organic carbon	18	Murphy (1974)
Noncarbonate carbon	18 – 44	Mueller, Mosley, and Pierce (1972)
Ammonium	0.7 – 2.7	McMullen, Faoro, and Morgan (1970)
	4.3	Moore (1977)
Nitrate	2.1 – 3.1	McMullen, Mosley, and Pierce (1972)
	1.3	Moore (1977)
	5 – 20[a]	Lundgren (1970)
	0.8 – 1.2	Shanty and Hemeon (1963)
	2.1 – 10.9	Williamson (1973)
Nitrate plus ammonium	9 – 18	Gordon and Bryan (1973)
Phosphorus	0.1	Murphy (1974)
ortho-Phosphorus	0.05	Murphy (1974)
Sulfate	7.5 – 13.1	Williamson (1973)
	9.9 – 22.2	McMullen, Faoro, and Morgan (1970)
	2 – 13[b]	Lundgren (1970)
	6.7 – 11.4	Shanty and Hemeon (1963)
Water soluble fraction	23 – 51[c]	Lundgren (1970)
	35	Estimated

[a] Average = 14.
[b] Average = 8.6.
[c] Average = 37.

are greater than the global background concentrations by a factor of 10 or more. For ammonia, carbon dioxide, and particulates, the ratio of urban to global background concentrations varies from less than 2 to 10.

For comparative purposes the federal ambient air quality criteria for the six air contaminants carbon monoxide, nitrogen oxides, sulfur dioxide, non-methane hydrocarbons, oxidants, and TSPs are summarized in Table 8.4. The non-methane hydrocarbon standard is not a standard in and of itself, and should be considered only if oxidant levels exceed the ambient air quality criterion. The values associated with the primary standards are based on health-related considerations, and values associated with the secondary standards are based on general welfare. Areas with air contaminants exceeding the federal criteria can be considered to be polluted areas, and those with air quality better than the criteria clean areas.

TABLE 8.2 CONCENTRATIONS OF SELECTED CONSTITUENTS OF PRECIPITATION

CONSTITUENT	CONCENTRATION mg l⁻¹	NOTES	REFERENCE
BOD	12–13		Loehr (1974)
COD	9–16		Loehr (1974)
Suspended solids	11.7–13.0		Loehr (1974)
Total nitrogen	0.73–1.3		Loehr (1974)
	0.64–0.65	Unfiltered	Caiazza, Hage, and Gallup (1978)
	0.24	Unfiltered (snow)	Caiazza, Hage, and Gallup (1978)
	0.48–0.50	Filtered	Caiazza, Hage, and Gallup (1978)
	0.18	Filtered (snow)	Caiazza, Hage, and Gallup (1978)
Nitrate nitrogen	0.14–1.1		Loehr (1974)
	0.18–0.61		Anlauf, Wiebe, and Stevens (1976)
	0.14–0.17	Unfiltered	Caiazza, Hage, and Gallup (1978)
Ammonium nitrogen	0.06–1.5		Loehr (1974)
	0.39–1.3		Anlauf, Wiebe, and Stevens (1976)
	0.32–0.38	Unfiltered	Caiazza, Hage, and Gallup (1978)
	0.08	Unfiltered (snow)	Caiazza, Hage, and Gallup (1978)
Nitrite nitrogen	0.002–0.003	Unfiltered	Caiazza, Hage, and Gallup (1978)
Organic nitrogen	0.26–0.33	Unfiltered	Caiazza, Hage, and Gallup (1978)
	0.16	Unfiltered (snow)	Caiazza, Hage, and Gallup (1978)
Inorganic nitrogen	0.7–0.9		Loehr (1974)
Sulfate	2.2–3.1		Loehr (1974)
	2.5–2.9		Caiazza, Hage, and Gallup (1978)
	5.7–11.5		Anlauf, Wiebe, and Stevens (1976)
Total phosphorus	0.034		Murphy (1974)
	0.02–0.04		Loehr (1974)
	0.022–0.125		Brezonik (1972)
ortho-Phosphorus	0.002–0.018		Brezonik (1972)
	0.017		Murphy (1974)
Total phosphate	0.008–0.04		Loehr (1974)
	0.055–0.082	Unfiltered	Caiazza, Hage, and Gallup (1978)

(Table continues)

TABLE 8.2 *(Continued)*

CONSTITUENT	CONCENTRATION mg l^{-1}	NOTES	REFERENCE
	0.039–0.055	Filtered	Caiazza, Hage, and Gallup (1978)
ortho-Phosphate	0.010–0.020	Unfiltered	Caiazza, Hage, and Gallup (1978)
	0.007–0.016	Filtered	Caiazza, Hage, and Gallup (1978)
Metaphosphate plus polyphosphate	0.023–0.036	Unfiltered	Caiazza, Hage, and Gallup (1978)
	0.020–0.033	Filtered	Caiazza, Hage, and Gallup (1978)
Organic phosphate	0.020–0.036	Unfiltered	Caiazza, Hage, and Gallup (1978)
	0.013–0.023	Filtered	Caiazza, Hage, and Gallup (1978)
pH	3.4–5.4	Canada (New Brunswick)	Anlauf, Wiebe, and Stevens (1976)
	2.1–5.0	Northeastern United States	Likens and Bormann (1974)

TABLE 8.3 TYPICAL CONCENTRATIONS OF ATMOSPHERIC CONTAMINANTS[a]

SUBSTANCE	CONCENTRATION (ppm)	
	GLOBAL BACKGROUND	URBAN AREAS
SO_2	0.0002	0.02–2.0
H_2S	0.0002	<0.1
CO	0.1	40–70
CO_2	320	400–600
NO_x	0.002–0.008	0.2
NO	0.0003–0.006	
NO_2	0.0005–0.004	
NH_3	0.001–0.02	0.02
Total hydrocarbons	<1–2	1–20
CH_4	1.5	2.5
non-CH_4	0.001	1.1
O_3	0.01–0.04	0.01–0.5
Particulates	10–20 $\mu g/m^3$	70–700 μ/m^3

[a] Data from Stern (1976) and Liptak (1974).

TABLE 8.4 SUMMARY OF FEDERAL AMBIENT AIR QUALITY STANDARDS

CONTAMINANT	AVERAGING TIME[a] (h)	PRIMARY STANDARD[b] (ppm)	PRIMARY STANDARD[b] (μg m^{-3})	SECONDARY STANDARD[b] (ppm)	SECONDARY STANDARD[b] (μg m^{-3})
TSPM	AGM[c]	—	75		60
	24	—	260		150
Sulfur dioxide	AAM[d]	0.03	80	None	None
	24	0.14	365	None	None
	3	None	None	0.5	1300
Carbon monoxide	8	9	10	Same as primary	
	1	35	40	Same as primary	
Non-methane hydrocarbons	3 (6 to 9 a.m.)	0.24	160	Same as primary	
Nitrogen dioxide	AAM	0.05	100	Same as primary	
Photochemical oxidants (O_3)	1	0.08	160	Same as primary	

[a] All standards with averaging time of 24 h or less are not to be exceeded more than once a year.
[b] At 25°C and 760 mm Hg.
[c] Annual geometric mean.
[d] Annual arithmetric mean.

The scope of this chapter does not include an evaluation of the potential effect on water quality of heavy metals or other trace substances in the atmosphere or estimates of the fluxes of these substances from the air to the land. Studies on the trace element content of TSPs (Moyers *et al.,* 1977; Peirson, Cawse, and Cambray, 1974; Peirson *et al.,* 1973) and of rainwater (Gatz and Dingle, 1971; Peirson, Cawse, and Cambray, 1974; Peirson *et al.,* 1973; Struempler, 1976) for selected elements are available elsewhere, and the methodology developed in this chapter could be applied to estimate regional load to the land for these substances.

IV. WET AND DRY DEPOSITION

Material is transferred from the atmosphere to land by both wet and dry deposition. Estimates of the fluxes of constituents that contribute to the water quality indices of interest are developed for both dry and wet weather conditions. The dry plus the wet deposition is an estimate of the total flux (deposition) of the constituent. These estimates are developed for a model atmosphere typical of a polluted urban area and for an urban area with clean air. The air quality for these two cases is exemplified by TSP concentrations of 100 and 50 μg m^{-3}, respectively. The other descriptors of air quality needed to develop the estimated fluxes are given in Section III, where the choice of particular values is discussed along with the method of estimating dry and wet deposition. Summaries of these example calculations are given in Tables 8.5–8.8.

A. Dry Deposition

1. Particulates

The dry deposition of particulates onto land is estimated in this example using a ratio of dustfall to total suspended particulates that was developed from air quality data collected in Chicago. These ratios can be developed for any given area in which dustfall and TSP data are available.

The ratio on an annual basis is

$$DF/TSPs = 14.0, \tag{1}$$

where DF is the dustfall in kilograms per hectare per year and TSPs the total suspended particulates in micrograms per cubic meter. The ratio when the TSPs are low, which occurs in the summer months, is

$$DF/TSPs = 8.40. \tag{2}$$

For polluted-air areas (TSPs = 100 μg/m^{-3}), the particulate flux would be approximately 1400 kg ha^{-1} yr^{-1}, using Eq. (1). For clean-air areas (TSPs = 50 μg m^{-3}), Eq. (2) can be used to estimate the particulate flux. The dry deposition of any particular constituent of the TSPs can be estimated from the product of the particulate flux and the weight fraction of the constituent of the TSPs. The results of these computations are summarized in Tables 8.5 and 8.6.

The weight fractions of the various constituents of TSPs are based on reported measurements, and the inorganic carbon fraction of TSPs can be estimated from rainwater analyses. Data for the dry deposition of phosphorus are based on the work of Murphy (1974). A total phosphorus content of 0.1% by weight was found for TSPs in Chicago, and one-half of this total was orthophosphorus.

The volatile fraction of TSPs is reported to be 25% (Stern, 1976). The total organic carbon as reported by Murphy (1974) and Friedlander (1973)

TABLE 8.5 DRY DEPOSITION IN CLEAN-AIR AREAS[a]

PARAMETER	FLUX (kg ha^{-1} yr^{-1})	
	PARTICULATE CONTRIBUTION	GASEOUS CONTRIBUTION
Total suspended particulates	420	
Volatile suspended particulates	110	
Total organic carbon	76	
Inorganic carbon (as C)		
Total	13	
Carbon dioxide	—	
Bicarbonate	13	
Phosphorus (as P)		
Total	0.42	
ortho-Phosphorus	0.21	
Organic phosphorus (by difference)	0.21	
Nitrogen (as N)		
Total inorganic nitrogen	3.8	16.2
Ammonium nitrogen	1.3	5.4
Nitrate nitrogen	2.5	10.8
Organic nitrogen	2.9	—
Total nitrogen	6.7	16.2
Total sulfur (as S)	11	16

[a] 50 μg m^{-3} total suspended particulates.

is approximately 18%. The (COD + BOD)/TOC ratio for a particlar petrochemical waste is 0.9 (Eckenfelder, 1970). This ratio is highly dependent on the chemical substance and is given here simply as an example. The organic compounds in the air are derived from petroleum and the combustion products of fossil fuels. In the absence of definitive information, the (COD + BOD)/TOC ratio of 0.9 is an educated guess.

The inorganic carbon fraction of TSPs is estimated to be 2–4% based on the bicarbonate and suspended solids content of rainwater (Fairbridge, 1972). The bicarbonate content is 5–6 ppm by weight in rain in the western United States. Because the pH of rain is usually less than 7, dissolved CO_2 and particulates are the major sources of inorganic carbon in rain.

In highly industrialized areas, the carbon dioxide content of air ranges from 300 to 600 ppm by volume. The corresponding carbon content of water (rain) saturated with CO_2 at atmospheric concentrations ranges from 0.12 to 0.24 ppm carbon by weight. A 5-ppm concentration of bicarbonates in rain corresponds to a 0.98-ppm carbon content. If the source of the

TABLE 8.6 DRY DEPOSITION IN POLLUTED-AIR AREAS[a]

PARAMETER	FLUX (kg ha^{-1} yr^{-1})	
	PARTICULATE CONTRIBUTION	GASEOUS CONTRIBUTION
Total suspended particulates	1400	
Volatile suspended particulates	350	
Total organic carbon	250	
Inorganic carbon (as C)		
Total	42	
Carbon dioxide	—	
Bicarbonate	42	
Phosphorus (as P)		
Total	1.4	
ortho-Phosphorus	0.7	
Organic phosphorus (by difference)	0.7	
Nitrogen (as N)		
Total inorganic nitrogen	12.6	32.4
Ammonium nitrogen	4.2	10.8
Nitrate nitrogen	8.4	21.6
Organic nitrogen	9.8	—
Total nitrogen	22.8	32.4
Total sulfur (as S)	38	33

[a] 100 μg m^{-3} total suspended particulates.

bicarbonates is rain with a suspended solids content of 20 to 40 ppm, then the inorganic carbon content of the TSPs is 2–4% carbon by weight. A value of 3% is used in calculations involving the inorganic carbon fluxes.

The ammonium, nitrite plus nitrate, and organic nitrogen contents of TSPs are typically 0.3%, 0.6%, and 0.7% nitrogen by weight. These values were developed from TSP compositions (Gordon and Bryan, 1973; Lee and Patterson, 1969; Stern, 1976) and rainwater analyses (Loehr, 1974). The ammonium and nitrate contents of TSPs in Washington, D. C., were reported to be 0.4 and 0.6% nitrogen by weight, respectively (Kowalczyk, Choquete, and Gordon, 1978). The sulfate content of TSPs is on the order of 2.7% sulfur by weight.

2. Gases

Estimates of the dry deposition of various atmospheric gases can be based on the concept of a deposition coefficient or velocity. This concept was introduced by Gregory (1945) in connection with the deposition of airborne spores and by Chamberlain (1953) in evaluating the deposition of radioactive materials from the atmosphere onto land. The deposition velocity is a fictional velocity; the product of the ground-level gaseous concentration and the deposition velocity equals the net flux of gas to a surface.

Deposition velocities for gases are generally determined from measuring the rate of decrease in the concentration of the gas over various types of surfaces. The deposition velocity of sulfur dioxide has been evaluated for different types of surfaces and conditions and ranges from 0.1 to 2.2 cm sec^{-1} (Garland, 1978). A commonly accepted value for the deposition velocity of SO_2 used in estimating the global sulfur budget is 0.8 cm sec^{-1}. Robinson and Robbins (1970) used a deposition velocity of 1 cm sec^{-1} for ammonia (obtained by analogy with SO_2) and 0.5 cm sec^{-1} for nitrogen dioxide in an analysis of the global nitrogen budget. The basis for these deposition velocities is unclear. The deposition velocity associated with the consumption of carbon monoxide by soil containing microorganisms ranges from 0.02 to 0.09 cm sec^{-1} (NRC Workshop, 1975).

The deposition velocity of a gas is dependent on factors such as the surface roughness of the terrain, atmospheric stability, temperature, the type of surface and presence or absence of moisture, the ease or difficulty with which the gas binds with the surface, and the physical and chemical properties of the gas. For water or soil surfaces, it appears that insoluble or slightly soluble gases have small deposition velocities, and soluble gases or gases that react chemically with the surface have significant deposition velocities.

Because there are numerous measurements of the deposition velocity of SO_2, the commonly accepted value of 0.8 cm sec^{-1} is used in estimating the gaseous dry deposition of SO_2 onto land. For urban areas a concentration of 0.01 ppm is typical of polluted-air areas and 0.005 ppm is typical of clean-air areas. The dry depositions of SO_2 onto land for these concentrations are 33 and 16 kg S ha^{-1} yr^{-1}.

The nominal atmospheric concentrations of NO, NO_2, and NH_3 are 0.03 ppm each in polluted-air areas and half that in clean-air areas. The TSP contents of NH_4^+ and NO_3^- are 0.3 and 0.6% by weight, respectively. An upper bound on the deposition velocity of NH_3 to surfaces similar to the particulate surfaces of TSPs can be estimated by attributing the NH_4^+ content of TSPs solely to NH_3 deposition and assuming a nominal retention time of 1 day for TSPs. Thus,

$$kvc = f(M/S), \tag{3}$$

where k is a unit conversion factor, v the deposition velocity in centimeters per second, c the gas concentration in parts per million by volume, f the weight fraction of NH_4^+ in TSPs in kilograms nitrogen per kilogram TSP per second, M the mass of TSPs in kilograms per particle, and S the particle surface area in hectares per particle.

For sulfur dioxide, $k = 1.306 \times 10^{-4}$ and $f = 0.027$ day^{-1}, or 3.125×10^{-7} sec^{-1}. Assuming a 4-μm diameter spherical particle with a density of 1 g cm^{-3}, $M/S = 6.67$ kg ha^{-1}, and an average SO$_2$ concentration of 0.00667 ppm gives a deposition velocity for SO$_2$ of 2.4 cm sec^{-1}. This average concentration is based on a concentration of 0.01 ppm at the surface that decreases parabolically to a concentration of 0.005 ppm at the top of the mixing layer. This estimate of deposition velocity is an upper bound, because not all of the sulfur content of TSPs is caused by gas transfer. Because the deposition velocity of SO$_2$ has been measured and generally accepted as 0.8 cm sec^{-1}, the sample calculation of deposition velocity is consistent with the measured value if one-third of the TSP sulfur content is caused by gas transfer. This permissible agreement between the calculated and measured deposition velocity for SO$_2$ is an indication that this method of estimating deposition velocity can give reasonable values for gases for which measured deposition velocities are not available.

In the case of ammonia or nitrogen dioxide, $k = 5.714 \times 10^{-5}$. To calculate S and M in this example, a representative particle diameter of 4 μm is used and particles are assumed spherical with a density of 1 g cm^{-3}. For ammonia $f = 0.003$ day^{-1}, and using an average concentration of 0.02 ppm NH$_3$ in the mixing layer of air, the deposition velocity of NH$_3$ is determined to be 0.2 cm sec^{-1} using Eq. (3). Similarly, the deposition velocity of NO$_2$ is estimated to be 0.4 cm sec^{-1} assuming an average mixing layer concentration of 0.02 ppm NO$_2$ in air, $f = 0.006$ day^{-1}, and the NO$_3^-$ content of TSPs is owing solely to NO$_2$ deposition. The contribution of NO to the nitrogen content of TSPs was considered negligible because NO is a slightly soluble gas. Of course, not all of the nitrogen content of TSPs can be attributed to gas deposition, and the surface characteristics of urban areas are different than TSP surfaces. However, there is some indication that over 90% of the inorganic nitrogen content of TSPs is attributable to gas transfer (Spicer and Schumacher, 1977).

It is also known that the ratio of NH$_4^+$ to NO$_3^-$ in rain varies greatly, from less than 0.1 to greater than 1.5 with a mean ratio of 1.1 being reported in a remote area of Norway (Marsh, 1978). On a global basis, the concentration of NH$_3$ is approximately twice that of NO$_2$. The rate of gas transfer to raindrops is expected to be greater than that to land or particulate surfaces because of more favorable transfer conditions. However, raindrop surfaces present equal accessibility for gas transfer of these two gases, and therefore imply that the deposition velocity for NH$_3$ is about one-half that for NO$_2$.

The dry depositions of NH$_3$ and NO$_2$ summarized in Tables 8.5 and 8.6 are based on a deposition velocity of 0.2 cm sec^{-1} for NH$_3$ and 0.4 cm sec^{-1} for NO$_2$. In a polluted urban atmosphere, the NH$_4^+$ and NO$_3^-$ fluxes, based on these velocities and on NH$_3$ and NO$_2$ ground-level concentrations of 0.03 ppm each in air, were estimated to be 10.8 and 21.6 kg N ha^{-1} yr^{-1}, respectively. For clean-air areas, ground-level concentrations of NH_3 and NO_2 of 0.015 ppm each were used.

The dry deposition of slightly soluble gases such as CO_2 and various hydrocarbons is largely unknown. It is unlikely that dry deposition of CO_2 is a large source of inorganic carbon to land. However, the transfer of CO_2 as well as O_2 to water surfaces is highly significant. The organic carbon flux due to dry deposition can be large, however.

3. Acidity

Acid vapors and acid nuclei exist in the atmosphere, and their concentrations are not insignificant. During pollution episodes the concentration of sulfuric acid aerosols can be 40 μg m^{-3} or greater, and mean ambient levels of nitric acid of 0.4 μg m^{-3} have been measured (Okita *et al.,* 1976). Hence, if the deposition velocity of the acid nuclei or vapor were comparable to that for SO_2, then the flux of hydrogen ion to land due to dry deposition of acids would be large or on the same order as the dry deposition of SO_2. There is, however, no information available to estimate accurately the deposition velocities of acid nuclei or vapors, nor the loss of acid nuclei to surfaces. Measurements of the dry deposition velocities of submicrometer particles containing principally lead (Dovland and Eliassen, 1976) and other submicrometer particles (Lodge, 1978) are reportedly comparable to the deposition velocity of sulfur dioxide.

B. Wet Deposition

1. Particulates

Rain and snow scavenge gases and particulates from the air. Rainout and snowout are associated with scavenging processes above the cloud level, and washout and sweepout are associated with processes below the cloud level. The concentration of various substances in precipitation is the result of all of these scavenging processes and is referred to as total wet deposition.

The estimates of the flux of various substances onto land are not differentiated as to rainout or snowout, because the information in the literature is inadequate for this purpose. Some information on the relative effectiveness of scavenging as a function of the precipitation intensity exists (Pasquill, 1962; Slade, 1968; Summers, 1970) and will therefore be discussed in this section and in the section on soluble gases.

The washout ratio (Englemann and Slinn, 1970; Inter-agency Committee, 1975), a dimensionless ratio of the concentration of a substance in rain to the concentration of the same substance in air, was used to estimate the flux of TSPs and associated chemical species to land. The flux of chemical species associated with the scavenging of gases will be discussed in the sections on slightly soluble and soluble gases. The washout ratio for TSPs is inferred from the washout ratio for radioactivity in Springfield, Illinois (Engelmann and Slinn, 1970) and for phosphorus in Chicago, Illinois (Murphy, 1974). The washout ratio for radioactivity was reported to be 490 and that for phosphorus 400. If concentration in rain is expressed as parts per million by weight and concentration in air is expressed as micrograms per cubic meter, then the washout ratio W is

$$W = (1180)(R/A_0), \qquad (4)$$

where W is a dimensionless number, R the concentration of a substance in rain in parts per million by weight, and A_0 the concentration of the same substance in air near the surface in micrograms per cubic meters. For Chicago Murphy reports a phosphorus content of 0.034 ppm for rain and snow, and 0.1% by weight for TSPs (all values expressed as concentrations of phosphorus). The soluble fraction of TSPs is expected to be 0.35, based on the major anions and associated cations (Stern, 1976), and the soluble fraction of dustfall computed from analysis of dustfall collected in Seattle, Washington is 0.39 (Johnson, Rossano, and Sylvester, 1966). It appears that the soluble fraction of suspended particulates and of deposited particulates is the same. Hence, the suspended solids content of rain can be expected to be

$$R_p = (0.65WA)/(1180). \qquad (5)$$

For $A = 100\,\mu m^{-3}$ and $W = 400$, R_p is estimated to be 22 ppm by weight; for $A = 50\,\mu m^{-3}$ and $W = 400$, R_p is 11 ppm by weight. The flux of insoluble material or suspended solids and volatile suspended solids to land on an annual basis is

$$F = 0.1Rp, \qquad (6)$$

where F is the flux in kilograms per hectare per year, p the precipitation in centimeters per year, the coefficient 0.1 is a unit adjustment factor. For each precipitation event, the flux can be adjusted on the basis of the precipitation intensity and the terminal settling velocity of the particulates suspended in air (Pasquill, 1972). The terminal settling velocity can be computed from Stoke's law. In applying Stoke's law, it should be noted that most atmospheric particles consisting of many primary particles have a density of approximately 1.0 g cm^{-3}, irrespective of the density of the primary particle (Bond and Straub, 1972). The particulate washout ratio for rain intensities less than 1 to 2 mm h^{-1} is essentially constant (Pasquill, 1962).

The source of phosphorus in rain is atmospheric particulates. For $A_0 = 100\,\mu g\ m^{-3}$ and $W = 400$, $R = 0.034$ ppm phosphorus. The *ortho*-phosphorus content of rain is reported to be one-half the total phosphorus content (Murphy, 1964), or 0.017 ppm.

The fluxes of total and *ortho*-phosphorus, total organic carbon, and various forms of nitrogen associated with suspended solids captured in rainwater are computed using the chemical composition of TSPs and Eqs. (4) and (6). The concentration of a particular substance in rain is calculated

TABLE 8.7   WET DEPOSITION IN CLEAN-AIR AREAS[a]

	FLUX (kg ha^{-1} yr^{-1})		CORRESPONDING RAINWATER CONCENTRATION (ppm)	
PARAMETER	PARTICULATE CONTRIBUTION	GASEOUS CONTRIBUTION	FROM CAPTURED PARTICULATES	FROM GAS TRANSFER
Total suspended particulates	81		11	
Volatile suspended particulates	20		2.8	
Total organic carbon	22		3.0	
Inorganic carbon (as C)				
Total	3.8	1.8		
Carbon dioxide	—	1.8		0.24
Carbonate	3.8		0.51	
Phosphorus (as P)				
Total	0.12		0.017	
ortho-Phosphorus	0.06		0.0085	
Organic phosphorus (by difference)	0.06		0.0085	
Nitrogen (as N)				
Total inorganic nitrogen	1.1	18.5	0.15	2.5
Ammonium nitrogen	0.37	11.8	0.05	1.6
Nitrate nitrogen	0.74	6.7	0.10	0.9
Organic nitrogen	0.81	—	0.11	—
Total nitrogen	1.9	18.5	0.26	—
Total sulfur (as S)	3.4	4.1	0.46	0.6
Hydrogen ion[b]				
pH 5.8	12		1.6 × 10^{-6}	
pH 5.0	74		10^{-5}	

[a] 50 μg m^{-3} total suspended particulates.

[b] For hydrogen ion flux is measured in equivalents per hectare per year, concentration in equivalents per liter.

using Eq. (5), in which the appropriate fraction, corresponding to the content of the substance in TSPs, is used. The fluxes of suspended and volatile solids are based on Eqs. (5) and (6). On a precipitation event basis, the phosphorus flux can be adjusted according to the load of suspended solids. Estimates based on Eqs. (4)–(6) and an annual precipitation rate of 74 cm yr^{-1} are summarized in Tables 8.7 and 8.8.

TABLE 8.8 WET DEPOSITION IN POLLUTED-AIR AREAS[a]

PARAMETER	FLUX (kg ha^{-1} yr^{-1})		CORRESPONDING RAINWATER CONCENTRATION (ppm)	
	PARTICULATE CONTRIBUTION	GASEOUS CONTRIBUTION	FROM CAPTURED PARTICULATES	FROM GAS TRANSFER
Total suspended particulates	162		22	
Volatile suspended particulates	40		5.5	
Total organic carbon	45		6.1	
Inorganic carbon (as C)				
Total	7.5	1.8		
Carbon dioxide	—	1.8		0.24
Carbonate	7.5		1.02	
Phosphorus (as P)				
Total	0.25		0.034	
ortho-Phosphorus	0.13		0.017	
Organic phosphorus (by difference)	0.12		0.017	
Nitrogen (as N)				
Total inorganic nitrogen	2.2	37	0.30	5.0
Ammonium	0.74	23.7	0.10	3.2
Nitrate	1.5	13.3	0.20	1.8
Organic nitrogen	1.8	—	0.24	—
Total nitrogen	4.0	37	0.54	—
Total sulfur (as S)	6.8	8.1	0.91	1.1
Hydrogen ion[b]				
pH 5.8	12		1.6×10^{-6}	
pH 5.0	74		10^{-5}	

[a] 100 μg m^{-3} total suspended particulates.
[b] For hydrogen ion flux as measured in equivalents per hectare per year, concentration in equivalents per liter.

2. Acidity

The acidity and pH of rain have received a great deal of attention recently (Likens, 1976). A pH of 5.8 is typical for rain in northeastern Illinois. In the northeastern United States, the annual mean rainwater pH is approximately 4, and for individual precipitation events the pH ranges from 2.1 to 5.0 (Likens, 1974).

A pH of 5.7 is in equilibrium with an atmospheric concentration of 330 ppm CO_2. For 600 ppm CO_2 in air, the equilibrium pH in water is 5.5. A pH different from the pH in equilibrium with CO_2 is due to influences of other acidic and basic gases, notably ammonia, sulfur dioxide, and nitrogen dioxide, and possibly due to equilibrium not being reached in the time that the rain is in contact with the atmosphere. The flux of hydrogen ion onto land from precipitation can be estimated using a revised form of Eq. (6) in which a unit conversion factor of 10^2 is used instead of 0.1. Of course, the pH of rainwater after contact with land is influenced by many chemical equilibria and cannot be treated so simply. The potential impact of acidity on a body of water depends on its alkalinity and buffering capacity.

Although most attention is currently focused on acid rain, alkaline snow with a pH of 9 has been reported recently in an area containing a large cement plant (Allam and Jonasson, 1978).

3. Slightly Soluble Gases

The gases nitric oxide (NO), nitrous oxide (N_2O), carbon monoxide (CO), carbon dioxide (CO_2), methane (CH_4), oxygen (O_2), acetylene (C_2H_2), and other hydrocarbons can be treated as insoluble or slightly soluble gases in water at the prevailing atmospheric concentrations. Table 8.9 gives the saturated concentrations of these gases in water corresponding to generally prevailing atmospheric concentrations, except in the case of acetylene. Data on oxygen is included to check that the saturated values are correct. The calculations are based on Henry's law:

$$P_g = Hx, \qquad (7)$$

where P_g is the partial pressure of the gas in air measured in atmospheres, H is Henry's law constant (atm^{-1}), and x is the mole fraction of the gas in water. The gaseous contribution to the wet deposition of CO_2 is owing to the transfer of CO_2 into the rainwater. In our example a rainwater concentration of 0.24 ppm for CO_2 was used. The wet deposition of CO_2 was then estimated using this rainwater concentration and Eq. (6).

Acetylene is included as an example of a relatively soluble hydrocarbon. The saturated concentrations in water (Table 8.9) show that the gaseous

TABLE 8.9 SATURATED CONCENTRATIONS OF SLIGHTLY SOLUBLE GASES IN WATER[a]

GAS	HENRY'S LAW CONSTANT $(atm^{-1}$ AT $25°C)$	SATURATED CONCENTRATION IN WATER (RATIOS BY WEIGHT)
NO (as N)	2.87×10^4	0.27 ppt
N_2O (as N)	2.25×10^3	0.17 ppb
Acetylene (as C)	1.33×10^3	1.0 ppb
CO_2 (as C)	1.64×10^3	0.24 ppm
CO (as C)	5.80×10^4	0.12 ppb
O_2	4.38×10^4	8.4 ppm
CH_4 (as C)	4.13×10^4	0.033 ppb
NH_3 (as N)[b]		0.011 ppm
NH_4 (as N)		
pH 8.0		0.18 ppm
pH 7.0		1.8 ppm
pH 6.0		18 ppm
pH 5.0		180 ppm
SO_2 (as S)[c]		3.0 ppb
HSO_3 (as S), pH 4.4		1.3 ppm
HCO_3 (as C)		
pH 8		11 ppm
pH 7		1.1 ppm
pH 6		0.11 ppm
HSO_3 (as S)		
pH 7		520 ppm
pH 6		52 ppm
pH 5		5.2 ppm

[a] Saturated concentrations are tabulated for the following concentration of gases in air at 20 to 25°C and 1 atm pressure (all ratios by volume): 2 ppm CH_4, 0.01 ppm NH_3, 0.01 ppm NO_2, 0.02 ppm SO_2, 0.01 ppm NO, 600 ppm CO_2, 10 ppm CO, 0.25 ppm N_2O, 1.0 ppm C_2H_2, 20.8% O_2.

[b] The solubility is 2 parts NH_3 per 100 parts water at 20°C, and the NH_3 partial pressure is 12 mm Hg.

[c] The solubility is 0.02 parts SO_2 per 100 parts water at 20°C, and the SO_2 partial pressure is 0.5 mm Hg.

contributions of carbon and nitrogen due to the wet deposition of slightly soluble gases are negligible.

4. Soluble Gases

The soluble gases of interest are ammonia (NH_3), sulfur dioxide (SO_2), and nitrogen dioxide (NO_2). The ammonium ion (NH_4^+) concentration in water is dependent on the partial pressure of NH_3 in air and the pH, in accordance with the hydrolysis of NH_3 in water:

$$K = [NH_4^+][OH^-]/[NH_3]. \tag{8}$$

At $25\,^\circ C$ $K = 1.65 \times 10^{-5}$ with the concentrations in moles per liter. The concentration of NH_3 can be calculated from Eq. (7). Ammonium ion concentrations for different pH values and an NH_3 concentration of 0.01 ppm in air are given in Table 8.9. The ammonium ion concentration of rain generally ranges from less than 0.1 ppm to several parts per million. In Urbana, Illinois, for example, the ammonium ion concentration in rain was found to be 0.1 ppm nitrogen (Fairbridge, 1972). Waite and Greenfield (1975) reported values in excess of 3.5 ppm nitrogen for Miami, Florida. It is therefore apparent that rain is usually not saturated with ammonia.

Similarly, the bisulfite ion (HSO_3^-) concentration in rain is governed by pH and the partial pressure of SO_2 in air, in accordance with the hydrolysis of SO_2 in water and provided that oxidation of the bisulfite does not take place. It is well known that oxidation *does* take place and the concentrations of sulfate and similar ions in rain are governed by gas transfer of SO_2 into rainwater. The bisulfite content of water in equilibrium with SO_2 in air and in the absence of oxidation is given in Table 8.9. The rain content of sulfur-containing ions ranges from several to 20 ppm (Englemann, 1970). It is apparent that sulfur-containing ions in rain up to several parts per million can be accounted for without involving oxidation of the sulfite ion. However, high concentrations of sulfur-containing ions in rain at pH values less than 6 can be explained only by considering oxidation.

The sulfate content of rain is based on gas transfer of SO_2 into rainwater and the rapid oxidation of sulfite to sulfate. This case can be thought of simply as gas transfer of SO_2 into water whose equilibrium with the liquid concentration of SO_2 is zero. However, wet deposition is not necessarily dominated by simple gas transfer, for example, rainout can be comparable or larger than washout in the case of SO_2.

The washout coefficient for gases accounts for both rainout and washout and is defined as (Slade, 1968)

$$A = A_0 e^{-\gamma t}, \tag{9}$$

where A is the concentration of SO_2 in air after a rain of duration t, A_0 the initial concentration of SO_2 in air prior to rain, γ the washout coefficient per hour, t the duration of the rain in hours, and $1 - e^{-\gamma t}$ the fraction of the atmospheric content of SO_2 prior to rain that is now associated with the rainwater. Values of γ/D, where D is the diffusivity of the gas in air, are summarized by Slade (1968) for soluble gases. For SO_2 and a rain intensity of 0.8 mm h^{-1}, γ/D ranges from 2×10^{-4} to 7×10^{-4} cm^{-2}, depending upon the raindrop size spectrum. For Argonne, Illinois, the median rain intensity and duration are 0.8 mm h^{-1} and 3 h, respectively (Moses and Bogner, 1967).

Starting with a SO_2 concentration in air of 0.01 ppm by volume at the surface and a parabolic decrease of SO_2 to 0.005 ppm at 1500 m above the surface, the SO_2 content of a 1500-m column of air over a surface area of 1 cm^2 is 0.0013 mg sulfur. The value of γ for SO_2 is 2.16×10^{-5} sec^{-1}, and γ/D is 2×10^{-4} cm^{-2}; therefore $A/A_0 = 0.2$. The value of A/A_0 ranges from 0.2 to 0.6 (Summers, 1970).

For rain intensities different than the reference intensity of 0.8 mm h^{-1} γ/D increases with intensity (Slade, 1968). At intensities of 0.08, 0.8, 8.0, and 80 mm h^{-1}, the corresponding values of γ/D are 0.6×10^{-4}, 2×10^{-4}, 20×10^{-4}, and 100×10^{-4} cm^{-2}. An increase in γ/D is accompanied by an increase in the amount of precipitation, and the net result is a decrease in the rainwater concentrations of substances washed out by rain.

The rainwater concentration for a soluble gas can be estimated by

$$R_g = 10^4(1 - A/A_0)m/it, \qquad (10)$$

where i is the rainfall intensity and m the mass (in milligrams) of trace gas, for example, SO_2, in a 1500-m air column above a 1-cm^2 surface area. We also have

$$m = (8.16 \times 10^{-3})M_w(a_0 - a_e), \qquad (11)$$

where M_w is the molecular weight of the element under consideration in grams, for example, M is 32 g for sulfur, a_0 the concentration of SO_2 in air at the surface in parts per million by volume as SO_2, and a_e the concentration of SO_2 in air at 1500 m above the surface in parts per million by volume as SO_2. For values of A/A_0 close to 1, A/A_0 in Eq. (10) can be approximated by $1 - \gamma t$.

The dissolution of NO_2 into water leads to decomposition through disproportionation. Essentially all of the NO_2 transferred to water ends up as nitrate. The estimation of the NO_3^- concentration in rain is made in the same manner as discussed for SO_2.

Using NO_2 concentrations of 0.03 and 0.015 ppm by volume as NO_2 at the surface and at 1500 m above the surface, respectively, $m = 1.71 \times 10^{-3}$ mg nitrogen. For $i = 0.8$ mm h^{-1}, $t = 3$ h, and $\gamma/D = 2 \times 10^{-4}$ as before, an estimate of the concentration of nitrate nitrogen in rain can be made using Eqs. (9) and (10). The diffusivity of NO_2 in air is 0.133 cm^2 sec^{-1}. The washout coefficient γ is then 2.66×10^{-5} sec^{-1}, A/A_0 is 0.75 [from Eq. (9)], and R_g is 1.8 ppm nitrogen [Eq. (11)].

The estimate of wet deposition of SO_2 is made using $a_0 = 0.01$ ppm and $a_e = 0.005$ pm by volume as SO_2. The value of m is 1.31×10^{-3} mg sulfur and $R_g = 1.1$ ppm sulfur.

In the case of ammonia, the diffusivity is 0.280 cm^2 sec^{-1}. Using $a_0 =$ 0.03 and $a_e = 0.015$ ppm as NH_3 and otherwise similar conditions, $R_g = 3.2$ ppm nitrogen. The sum of NH_4 and NO_3 nitrogen in rain for the example developed here is 3.2 +1.8 = 5.0 ppm, which is consistent with the composition of rainwater. However, the NH_4 nitrogen is somewhat high and the NO_3 nitrogen is low, suggesting that oxidation takes place. The measurements of NH_4 and NO_3 nitrogen in rainwater are often not made until several days after collection of the rainwater, thus permitting time for oxidation to occur. This interpretation is also consistent with the fact that storm runoff is very low in NH_4 nitrogen, although the absorption of NH_4 by land from runoff waters may also be responsible for the low ammonium ion concentration.

5. Total Wet Deposition

For an estimation of the total wet deposition, the gaseous components transferred from the air to rain must be added to the constituents transferred to rain by the TSPs. For sulfates the wet deposition contribution computed by Eq. (4) must be added to the wet deposition computed by Eq. (10). As an example, $A_0 = 100$ μg m^{-3}, $W = 400$, and R_p from Eq. (4) is 33.9 ppm suspended solids. The sulfur content is 0.027 $\times$ 33.9 or 0.91 ppm. The corresponding R_g from gaseous transfer is 1.1 ppm, and the sum of these two contributions gives 2.01 ppm sulfur. The wet deposition flux of sulfur-containing ions onto land is computed using Eq. (6) and $p = 74$ cm yr^{-1} and is 14.9 kg ha^{-1} yr^{-1}.

Continuing with the same example, the ammonium ion content of TSPs is 0.3%. Hence, $R_p = 0.003 \times 33.9 = 0.10$ ppm nitrogen is the TSP contribution, and the gaseous contribution R_g is 3.2 ppm nitrogen, giving a total of 3.3 ppm NH_4 nitrogen. The flux is then 24.4 kg ha^{-1} yr^{-1}, using Eq. (6).

Similarly, the nitrate nitrogen content is $R_p = 0.006 \times 33.9 = 0.20$ ppm from the TSP contribution. The gaseous contribution is $R_g = 1.8$ ppm, giving a total rainwater concentration of 2.0 ppm nitrogen and a flux of 14.8 kg ha^{-1} yr^{-1}. There is also an organic nitrogen flux associated with the suspended particulates captured in the rainwater and, based on an organic nitrogen content of 0.7% in TSPs, this is 1.8 kg N ha^{-1} yr^{-1}. The total wet deposition of nitrogen compounds is 41 kg N ha^{-1} yr^{-1}.

V. TOTAL DEPOSITION

The total wet and dry deposition for suspended particulates is large. Total depositions of 500 and 1560 kg ha^{-1} yr^{-1} were estimated for clean-air and

TABLE 8.10   TOTAL DEPOSITION IN CLEAN-AIR AREAS[a]

PARAMETER	FLUX (kg ha⁻¹ yr⁻¹)	
	PARTICULATE CONTRIBUTION	GASEOUS CONTRIBUTION
Total suspended particulates	500	
Volatile suspended particulates	130	
Total organic carbon	98	
Inorganic carbon (as C)		
Total	16.8	1.8
Carbon dioxide	—	1.8
Carbonate	16.8	
Phosphorus (as P)		
Total	0.54	
ortho-Phosphorus	0.27	
Organic phosphorus (by difference)	0.27	
Nitrogen (as N)		
Total inorganic nitrogen	4.9	34.7
Ammonium	1.67	17.2
Nitrate	3.24	17.5
Organic nitrogen	3.71	—
Total nitrogen	8.6	34.7
Total sulfur (as S)	14.4	20.1
Hydrogen ion[b]		
pH 5.8	12	
pH 5.0	74	

[a] 50 μg m⁻³ total suspended particulates.
[b] For hydrogen ion flux is measured in equivalents per hectare per year.

polluted-air areas, respectively. Estimates of total depositions of the nutrients nitrogen and phosphorus are 43.3 kg N ha⁻¹ yr⁻¹ and 0.54 kg P ha⁻¹ yr⁻¹ in clean-air areas and 95.8 kg N ha⁻¹ yr⁻¹ and 1.6 kg P ha⁻¹ yr⁻¹ in polluted-air areas, respectively. Total carbon fluxes to land are estimated at 98 and 292 kg ha⁻¹ yr⁻¹. Summaries of the total deposition in clean-air and polluted-air areas are given in Tables 8.10 and 8.11.

The potential impact of the total atmospheric deposition of a substance on water quality depends on the particular situation. For clean bodies of water and in the absence of major pollution sources, the atmosphere can be a significant source of contaminants. For large streams with many sources of pollution, the contribution of atmospheric contaminants to the deterioration of water quality is probably small.

TABLE 8.11   TOTAL DEPOSITION IN POLLUTED-AIR
AREAS[a]

PARAMETER	FLUX (kg ha^{-1} yr^{-1})	
	PARTICULATE CONTRIBUTION	GASEOUS CONTRIBUTION
Total suspended particulates	1560	
Volatile suspended particulates	390	
Total organic carbon	295	
Inorganic carbon (as C)		
Total	49.5	1.8
Carbon dioxide	—	1.8
Carbonate	49.5	
Phosphorus (as P)		
Total	1.65	
ortho-Phosphorus	0.83	
Organic phosphorus (by difference)	0.82	
Nitrogen (as N)		
Total inorganic nitrogen	14.8	69.4
Ammonium	4.9	34.5
Nitrate	9.9	34.9
Organic nitrogen	11.6	—
Total nitrogen	26.4	69.4
Total sulfur (as S)	44.8	41.1
Hydrogen ion[b]		
pH 5.8	12	
pH 5.0	74	

[a] 100 μg m^{-3} total suspended particulates.
[b] For hydrogen ion flux is measured in equivalents per hectare per year.

VI. CONCLUSION

A method for estimating the fluxes onto land of substances in the air that have a potential impact on a selected set of water quality indices of interest has been developed, along with sample calculations. The actual contribution of air contaminants to the deterioration of water quality depends on the fraction of the air to land flux transferred to water, the rate at which this fraction reaches a body of water, and the capacity of the body of water to assimilate the substance in question. The burden of contaminants to a body of water is generally less than the wet deposition of flux onto the land

and much less than the dry deposition. It appears that, on an annual basis, the soil or land and its coverings retain a great deal of the material deposited on it and, indeed, remove material from rainwater flowing over it.

The soil has a large capacity for retaining material deposited on or coming into contact with it. Microorganisms and other substances in soil are able to transform the retained material, with reaction products subsequently released to the atmosphere. The absorptive or retentive capacity of soil for various cations and anions is well recognized, and some types of soil are known to have large absorptive capacities, for example, for ammonium ion. The uptake of sulfur dioxide by vegetation or soil results in its transformation to sulfate and subsequently to hydrogen sulfide, which is released to the atmosphere. Retention and transformation of materials reaching the land mean that the most likely contributions of air contaminants to the deterioration of water quality are associated with wet deposition and materials on land that are easily transported by stormwater runoff. The scavenging of suspended particulates and trace gases in air by precipitation and dustfall, then, is expected to be the significant mode of contribution.

Sample calculations of the total flux from air to land for various substances show that fluxes in urban areas are an order of magnitude larger than fluxes on a global basis (Singer, 1970). This is simply a reflection of the fact that urban concentrations of air contaminants are much larger than the average global concentrations. Clean-air and polluted-air areas have a similar potential for contributing to water pollution, because air contaminant concentrations for these areas differ by a factor of only two. In urban areas dry deposition accounts for 51 to 90% of the total deposition for the constituents considered, except for ammonium nitrogen, for which dry deposition is approximately 36–38% of total deposition. If dry gaseous deposition is ignored, because it contributes very little to the deterioration of water quality, then the dry deposition flux is 25–90% of the total flux. Once again, ammonium nitrogen differs in that its wet deposition is about 85% of its total deposition.

In the case of total inorganic nitrogen, stream loads from storm runoff are on the order of a few to 10 kg N $ha^{-1} yr^{-1}$. The total flux from air to land for polluted-air areas, without considering dry gaseous deposition, is 51.6 kg N $ha^{-1} yr^{-1}$; the wet deposition is 39 kg $ha^{-1} yr^{-1}$ and the dry deposition associated with particulates is 12.6 kg $ha^{-1} yr^{-1}$. (Recall that the inorganic nitrogen associated with particulates is water soluble.) Consequently, perhaps 20–50% of the total deposition will reach surface water systems.

The fraction of atmospheric flux that reaches a body of water has not received much attention. Also, dustfall measures the flux of TSPs to a horizontal surface without considering resuspension or reentrainment. Net dustfall is usually not measured, and the dustfall measurement has been abandoned in recent years. Estimates of the fluxes of dry particulates and associated materials are probably overestimated, if based on dustfall data. Differences in wet deposition fluxes due to scavenging by rain versus scavenging by snow are largely unknown, except in the case of phosphorus associated with particulates, for which scavenging by rain and by snow were determined to be the same (Murphy, 1974).

The stream loading of inorganic nitrogen due to domestic sewage, based on a flow of 100 gal person^{-1} day^{-1}, 10 persons acre^{-1}, and 6 mg N l^{-1} sewage, is on the order of 20 kg N ha^{-1} yr^{-1}. For areas with advanced waste treatment, the 5-day BOD and the concentrations of suspended solids and total inorganic nitrogen in the effluent are a few parts per million each. Phosphorus concentrations are usually less than 1 ppm, and total inorganic nitrogen is less than 1 ppm if a denitrification process is used. For urban areas with advanced waste treatment, the regional loads from domestic sewage would therefore be 13 kg BOD ha^{-1} yr^{-1}, 13 kg suspended solids ha^{-1} yr^{-1}, 3–20 kg N ha^{-1} yr^{-1}, and 3.4 kg P ha^{-1} yr^{-1}.

A significant fraction of the flux from air to land, except for dry gaseous deposition and ammonium nitrogen, reaches the water. The atmospheric dry deposition of phosphorus is 1.4 kg ha^{-1} yr^{-1} and the wet deposition is 0.25 kg ha^{-1} yr^{-1}. It thus appears that if one-half of the total flux reaches a waterway, the atmospheric contribution of phosphorus would be about 25% of the domestic sewage contribution. In the case of inorganic nitrogen, the atmospheric contribution would be as large or larger than the domestic sewage contribution, and the atmospheric contribution of suspended solids would be many times the domestic sewage contribution.

The impact of a source of contaminants on water quality and the significance of this impact frequently depend on the *change* in concentration of a contaminant in the water environment rather than on the concentration itself or on the stream loading. For a clean body of water, relatively small loads would result in a large change in concentration with significant adverse impacts, whereas moderately polluted waters can receive similar loads without adverse impacts.

The total organic carbon content of materials of atmospheric origin (TSPs) and of materials commonly associated with water pollution are not equivalent. The TOC measurement of water pollution is intended to measure material that is readily biodegradable. The TOC content of TSPs simply measures carbon residuals from combustion that are probably not readily biodegradable; therefore, BOD measurements are more useful in this case.

The potential significance of certain modes of contribution can be masked by the use of annual fluxes. There are both temporal and spatial variations in air and water quality indices. Seasonal effects can be important because both air and water quality parameters are known to show distinct and large seasonal variations. Hence, fluxes due to the dry and/or wet deposition of various materials can differ considerably from the annual flux, depending on the time of year. Snow-covered land can facilitate or deter dry deposition processes. Materials deposited onto snow surfaces and stored there can enter the water environment rapidly during snowmelt. In Scandanavia, for example, large decreases in the pH of lakes are observed in the spring. Oden (1976) attributes this decrease to the acidity of the melting snow cover. The deposition velocity of SO_2 over snow is much less than the nominal value owing to the extreme stability of the atmosphere associated with the snow cover (Dorland and Eliassen, 1976).

Our sample calculations are for urban areas with clean and polluted air. If sufficient air quality data are available, the urban area under study can be divided into zones based on isopleths of air quality and atmospheric contributions for each zone computed separately. Also, each precipitation event can be considered separately and its impact on a body of water evaluated taking into consideration the hydrology of the particular situation.

The fluxes of heavy metals and pesticides from the air to land were not evaluated here owing to the set of water quality indices selected. However, the approach presented can be applied to evaluate the potential contribution of these substances to water pollution. In these applications one should be aware that heavy metals tend to be associated with small particles and that pesticides can exist as small particulates and/or vapors.

REFERENCES

Allan, R. J., and Jonasson, I. R. (1978). Alkaline snowfalls in Ottawa and Winnipeg, Canada. *Atmos. Environ.* **12**, 1169.

Anlauf, K. G., Wiebe, H. A., and Stevens, R. D. S. (1976). Some measurements of air quality and precipitation in Saint John, N. B. *Water, Air, Soil Pollut.* **5**, 507.

Bond, R. T., and Straub, C. P., eds. "Handbook of Environmental Control, Vol. I: Air Pollution." CRC Press, The Chemical Rubber Co., Cleveland, Ohio.

Bormann, F. H., Likens, G. E., Fisher, D. W., and Pierce, R. S. (1968). Nutrient loss accelerated by clear-cutting of a forest ecosystem. *Science* **159**, 882.

Brezonik, P. L. (1972). Nitrogen: sources and transformations in natural waters. *In* "Nutrients in Natural Waters." H. E. Allen and J. R. Kramer, eds. Wiley (Interscience), New York.

Caiazza, R., Hage, K.D., and Gallup, D. (1978). Wet and dry deposition of nutrients in Central Alberta. *Water, Air, Soil Pollut.* **9**, 309.

Chamberlain, A. C. (1953). "Aspects of Travel and Deposition of Aerosol and Vapour Clouds." AERE Report, HP/R 1261, H. M. S. O., London, United Kingdom.

Chamberlain, A. C. (1960). Aspects of the deposition of radioactive and other gases and particles. *Int. J. Air Water Pollut.* **3**, 63.

Davies, T. D. (1976). Precipitation scavenging of sulfur dioxide in an industrial area. *Atmos. Environ.* **10**, 879.

Dovland, H., and Eliassen, A. (1976). Dry deposition on a snow surface. *Atmos. Environ.* **10**, 783.

Eckenfelder, W. W., Jr. (1970). "Water Quality Engineering for Practicing Engineers." Barnes and Noble, New York.

Englemann, R. J. (1970). Scavenging prediction using ratios of concentrations in air and precipitation. *In* "Precipitation Scavenging" (R. J. Englemann and G. N. Slinn, eds.). Proceedings of Symposium, Richland, Washington, USAEC, NTIS Conf. 700601.

Englemann, R. J., and Slinn, W. G. N., coordinators (1970). "Precipitation Scavenging," Proceedings of Symposium, Richland, Washington, U. S. Atomic Energy Commission, NTIS Conf. 700601.

Fairbridge, R. W., ed. (1972). "The Encyclopedia of Geochemistry and Environmental Sciences, Encyclopedia of Earth Sciences, Vol. IVA." Van Nostrand-Reinhold, Princeton, New Jersey.

Fisher, D. W., Gambell, A. W., Likens, G. E., and Bormann, F. H. (1968). Atmospheric contributions to water quality of streams in the Hubbard Brook Experimental Forest, New Hampshire. *Water Resour. Res.* **4**, 1115.

Fowler, D. (1978). Dry deposition of SO_2 on agricultural crops. *Atmos. Environ* **12**, 369.

Friendlander, S. K. (1973). Chemical element balances and identification of air pollution sources. *Environ. Sci. Technol.* **7**, 235.

Garland, J. A. (1978). Dry and wet removal of sulphur from the atmosphere. *Atmos. Environ.* **12**, 349.

Gatz, D. F. (1975). Pollutant aerosol deposition into southern Lake Michigan. *Water, Air, Soil Pollut.* **5**, 239.

Gatz, D. F., and Dingle, A. N. (1971). Trace substances in rain water: concentration variation during convective rains, and their interpretation. *Tellus* **23**, 14.

Gordon, R. J., and Bryan, R. J. (1973). Ammonium nitrate in airborne particles in Los Angeles. *Env. Sci. Technol.* **7**, 645.

Gregory, P. H. (1945). The dispersion of air-borne spores. *Trans. Br. Mycol. Soc.* **28**, 26.

Harmeson, R. H., Sollo, F. W., Jr., and Larson, T. E. (1971). The nitrate situation in Illinois. *J. Am. Water Works Assoc.* **63**, 303.

Hasler, A. D. (1970). Man-induced eutrophication of lakes. *In* "Global Effects of Environmental Pollution" (S. F. Singer, ed.). Springer-Verlag, Berlin and New York.

Huff, F. A. (1976). Relation between atmospheric pollution, precipitation and streamwater quality near a large urban–industrial complex. *Water Res.* **10**, 945.

Inter-agency Committee on Marine Science and Engineering of the Federal Council for Science and Technology (1975). "Proceedings of the Second Federal Conference on the Great Lakes." Public Information Service Office of the Great Lakes Basin Commission.

Johnson, R. E., Rossano, A. T., Jr., and Sylvester, R. O. (1966). Dustfall as a source of water quality impairment. *J. San. Eng. Div., Am. Soc. Civ. Eng.* **92**(SA1), 245.

Kowalczyk, G. S., Choquette, C. E., and Gordon, G. E. (1978). Chemical element balances and identification of air pollution sources in Washington, D. C." *Atmos. Environ.* **12**, 1143.

Lee, R. E., and Patterson, R. K. (1969). Size determination of atmospheric phosphate, nitrate, chloride and ammonium particulate in several areas. *Atmos. Environ.* **3**, 249.

Likens, G. (1976). Acid precipitation. Special Report in *Chem. Eng. News* **54**, 29.

Likens, G. E., and Bormann, F. H. (1974). Acid rain: a serious regional environmental problem. *Science* **184**, 1176.

Likens, G. E., Bormann, F. H., Johnson, N. M., and Pierce, R. S. (1967). The calcium, magnesium, potassium and sodium budgets for a small forested ecosystem. *Ecology* **48**, 772.

Liptak, B. G., ed. (1974). "Environmental Engineers' Handbook, Vol. 2: Air Pollution." Chilton Book Co., Radnor, Pennsylvania.

Lodge, J. P., Jr. (1978). An estimate of deposition velocities over water. *Atmos. Environ.* **12**, 973.

Loehr, R. C. (1974). Characteristics and comparative magnitude of nonpoint sources. *J. Water Pollut. Control Fed.* **46**, 1849.

Lundgren, D. A. (1970). Atmospheric aerosol composition and concentration as a function of particle size and time. *J. Air Pollut. Control Assoc.* **20**, 603.

McMahon, T. A., Denison, P. J., and Fleming, R. (1976). A long-distance air pollution transportation model incorporating washout and dry deposition components. *Atm. Environ.* **10**, 751.

McMullen, T. B., Faoro, R. B., and Morgan, G. B. (1970). Profiles of pollutant fractions in nonurban suspended particulate matter. *J. Air Pollut. Control Assoc.* **20**, 369.

Marsh, A. R. W. (1978). Sulphur and nitrogen contributions to the acidity of rain. *Atmos. Environ.* **12**, 401.

Mayer, R., and Ulrich, B. (1978). Input of atmospheric sulfur by dry and wet deposition to two Central European forest ecosystems. *Atmos. Environ.* **12**, 375.

Moore, H. (1977). The isotopic composition of ammonia, nitrogen dioxide and nitrate in the atmosphere. *Atmos. Environ.* **11**, 1239.

Moses, H., and Bogner, M. A. (1967). "Fifteen-Year Climatological Summary, Jan. 1, 1950 to Dec. 31, 1964." Argonne National Laboratory, ANL-7084, TID-4500, Argonne, Illinois.

Moyers, J. L., Ranweiler, L. E., Hopf, S. B., and Korte, N. E. (1977). Evaluation of particulate trace species in southwest desert atmosphere. *Environ. Sci. Technol.* **11**, 789.

Mueller, P. K., Mosley, R. W., and Pierce, L. B. (1972). Chemical composition of Pasadena aerosol by particle size and time of day. IV. Carbonate and noncarbonate carbon content. *In* "Aerosols and Atmospheric Chemistry" (G. M. Hidy, ed.) Academic Press, New York.

Murphy, T. J. (1974). "Sources of Phosphorus Inputs from the Atmosphere and Their Significance to Oligotrophic Lakes." University of Illinois at Urbana-Champaign, Water Resources Center, UILU-WRC-74-0092, Research Report No. 92.

Murphy, T. J. (1976). "Precipitation: A Significant Source of Phosphorus and PCB's to Lake Michigan." 10th ACS Great Lakes Regional Meeting, Evanston, Illinois.

Murphy, T. J., and Doskey, P. V. (1975). "Inputs of Phosphorus from Precipitation to Lake Michigan." Environmental Protection Agency Office of Research and Development, Environmental Research Laboratory, Duluth, Minnesota, EPA-600/3-75-005.

National Research Council Workshop on Tropospheric Transport of Pollutants to the Ocean Steering Committee (1975). "The Tropospheric Transport of Pollutants and Other Substances to the Ocean." National Academy of Sciences, Washington, D. C.

Oden, S. (1976). The acidity problem — An outline of concepts. *Water, Air, Soil Pollut.* **6**, 137.

Okita, T., Morimoto, S., Izawa, M., and Konno, S. Measurements of gaseous and particulate nitrates in the atmosphere. *Atmos. Environ.* **10**, 1085.

Pasquill, F. (1962). "Atmospheric Diffusion: The Dispersal of Windborne Material from Industrial and Other Sources." Van Nostrand-Reinhold, Princeton, New Jersey.

Peirson, D. H., Cawse, P. A., and Cambray, R. S. (1974). Chemical uniformity of airborne particulate material, and a maritime effect. *Nature* **251**, 675.

Peirson, D. H., Cawse, P. A., Salmon, L., and Cambray, R. S. (1973). Trace elements in the atmospheric environment. *Nature* **241**, 252.

Peyton, T., McIntosh, A., Anderson, V., and Yost, K. (1976). Aerial input of heavy metals into an aquatic ecosystem. *Water, Air, Soil Pollut.* **5,** 443.

Robinson, E., and Robbins, R. C. (1970). Gaseous atmospheric pollutants from urban and natural sources. *In* "Global Effects of Environmental Pollution" (S. F. Singer, ed.) Springer-Verlag, Berlin and New York.

Shanty, F., and Hemeon, W. C. L. (1963). The inhalability of outdoor dust in relation to air sampling networks. *J. Air Pollut. Control Assoc.* **13,** 211.

Sievering, H. (1976). Dry deposition loading of Lake Michigan by airborne particulate matter. *Water, Air, Soil Pollut.* **5,** 309.

Sievering, H., and Williams, A. C. (1975). "Potential Loading of Southern Lake Michigan by Dry Deposition." 2nd ICMSE Conference on the Great Lakes, Argonne National Laboratory, Argonne, Illinois.

Singer, S. F., ed. (1970). "Global Effects of Environmental Pollution." Springer-Verlag, Berlin and New York.

Skibin, D. (1973). Comment on water pollution in Lake Michigan from pollution aerosol fallout. *Water, Air, Soil Pollut.* **2,** 405.

Slade, D. H., ed. (1968). "Meteorology and Atomic Energy." U. S. Atomic Energy Commission TID-24190.

Spicer, C. W., and Schumacher, P. M. (1977). Interferences in sampling atmospheric particulate matter. *Atmos. Environ.* **11,** 873.

Stern, A. C., ed. (1976). "Air Pollution," vol. I, 3rd ed. Academic Press, New York.

Struempler, A. W. (1976). Trace metals in rain and snow during 1973 at Chadron, Nebraska. *Atmos. Environ.* **10,** 33.

Summers, P. W. (1970). Scavenging of SO_2 by convective storms. *In* "Precipitation Scavenging" (R. J. Englemann and G. N. Slinn, eds.). Proceedings of Symposium, Richland, Washington, U. S. Atomic Energy Commission, NTIS Conf. 700601.

Wagner, G. H., and Holloway, R. W. (1975). Sodium, potassium, calcium and magnesium content of northwest Arkansas rainwater in 1973. *J. Appl. Meteorol.* **14,** 578.

Waite, T. D., and Greenfield, L. J. (1975). Stormwater runoff characteristics and impact on urban waterways. Proceedings of the American Society of Civil Engineers Specialty Conference, University of Florida, Gainesville, Florida.

Williamson, S. J. (1973). "Fundamentals of Air Pollution." Addison-Wesley, Reading, Massachusetts.

Winchester, J. W., and Nifong, G. D. (1971). Water pollution in Lake Michigan by trace elements from pollution aerosol fallout. *Water, Air, Soil Pollut.* **1,** 50.

Chapter 9

Water Quality Modeling

by NEIL J. FREEMAN*

I. INTRODUCTION

Environmental systems besides those used for water and wastewater treatment must also be mathematically modeled to accomplish rational resource planning. Predictive results of water quality changes, for instance, due to modifications in watershed areas, can only be made if accurate, functional models exist to represent the situation in question. In these cases modelers attempt to dissect complex phenomena into understandable subsets. These subsets can then be investigated mathematically to observe their relative contribution to the whole. Environmental models are not only written to evaluate quality changes that may occur owing to human activity but also to provide a basis for economic analysis. For example, a model could be developed to represent the movement of phosphorus through a system (including biological and physical–chemical interactions). Utilizing this model, a community could evaluate its input to the phosphorus pool

* Department of Civil Engineering, University of Miami, Coral Gables, Florida.

through sewage discharges and surface runoff. A decision could then be made concerning the necessity for initiating a costly tertiary treatment process for the sewage, or perhaps for developing a system for the diversion of runoff waters. Thus, it is possible for engineers and administrators to evaluate the environmental significance of a project as well as the cost-benefit ratio.

It should be apparent to the reader that the system analysis approach is the framework used for the development of most environmental models. These mathematical models allow for the making of rational decisions concerning environmental modifications.

In an effort to describe the many states of a system, we are often faced with a tremendously complex task. To avoid attempting to establish a relation among all states of a process, we can investigate a single relation between one state and the next. This type of relation expresses how a certain state will develop in the near future and is based on what is usually referred to as the "law of causality," which allows us to describe mathematically natural processes in terms of infinitesimal increments of time and space. Relations among these derivatives and other functions are differential equations. Thus, differential equations express relationships between increments of certain quantities and the quantities themselves. From this discussion it appears that a differential equation is a good representation of the law of causality.

II. MATERIAL TRANSPORT

We shall now describe how chemicals or biological material within fluids are moved from one point to another. This movement is designated *material transport*. If the material moves with the fluid, the material transport is called *advection*. If the material moves within the fluid, owing to differences in concentration throughout the fluid, it is called *diffusion*. In any fluid system material can be transported by either or both of these processes.

A. Mass Transport Rates

A *homogeneous* fluid has a uniform composition at all points within its boundary. If the concentration of a chemical or biological material varies with location, concentration gradients result. Mathematically, these gradients are rates of change of concentration with respect to position within the fluid system.

Other types of gradients can exist within fluid systems and cause material transport. For example, gradients can be caused by differences in pressure or temperature. Gradients due to unequal forces acting throughout a fluid, such as the centrifugal force that acts on substances within a rotating fluid, can also cause material transport.

To represent and analyze the processes of advection and diffusion, we must define the variables that affect the concentration of materials within a fluid system. Let us first consider C_A, the concentration of material A, in moles per liter. The dependence of the concentration on location within the fluid system leads to the selection of the variables x, y, and z as the three spatial rectangular coordinates located with respect to a fixed reference frame. Any material transport process represents a change in location of substances over time, and therefore t will be used to represent the temporal variable. The four variables x, y, z, and t are sufficient to describe the concentration C_A and its changes within the fluid system. Therefore, in general we can write

$$C_A = C_A(x, y, z, t). \tag{1}$$

This notation indicates that C_A is a function of the four variables x, y, z, and t.

Most of the following discussion will be simplified by considering only concentrations dependent on one spatial coordinate, represented by

$$C_A = C_A(x, t). \tag{2}$$

In certain cases this simplification is justified by the symmetry of the physical situation and therefore results in no loss of generality. In other cases the development in terms of only x will be representative of the effect of the other spatial variables.

The variation of the concentration of material A with respect to the position variable x is written $\partial C_A / \partial x$, which denotes the partial derivative of C_A with respect to x and represents the rate of change of C_A in the x direction holding time fixed. The variation of C_A with respect to the time variable t is written $\partial C_A / \partial t$. This partial derivative represents the rate of change of C_A that occurs within a particular fixed volume element (holding x constant) through which fluid is flowing.

Finally, we introduce a special type of total time derivative that represents the change in C_A in a specific element of fluid that flows along with the surrounding fluid. This derivative is written

$$\frac{DC_A}{Dt} = \frac{\partial C_A}{\partial x} V_x + \frac{\partial C_A}{\partial t}, \tag{3}$$

where V_x is the speed of the specific element of the flow in the x direction. This derivative has been given various names in the literature, such as "material," "connective," "particle," and "derivative following the motion."

In three dimensions the total time derivative DC_A/Dt takes the form

$$\frac{DC_A}{Dt} = \left(\frac{\partial C_A}{\partial x} V_x + \frac{\partial C_A}{\partial y} V_y + \frac{\partial C_A}{\partial z} V_z \right) + \frac{\partial C_A}{\partial t}, \tag{4}$$

where the terms $\partial C_A/\partial y$, $\partial C_A/\partial z$, V_y, and V_z are analogous to the previously defined terms $\partial C_A/\partial x$ and V_x.

In a quiescent system, that is, one in which there is no flow, $V_x = V_y = V_z = 0$, and therefore

$$DC_A/Dt = \partial C_A/\partial t. \tag{5}$$

Recall that $\partial C_A/\partial t$ represents the change in concentration of material A within a specific fixed volume element and that DC_A/Dt represents the change in concentration within a specific element of fluid moving with the flow. When there is no flow, the specific volume element is continuously occupied by a specific element of fluid, and the two types of change in concentration are equivalent. In the case of zero flow, therefore, changes in the concentration C_A occur by molecular action only, which is called *diffusive* mass transport. This is also the type of material transport that one would observe if moving along with a specific element of fluid when the flow is nonzero.

B. Law of Continuity

The law of conservation of mass can be applied to a material A flowing through a *stationary* volume element to yield a continuity equation in terms of the mass density ρ_A in milligrams per liter. [We have used the molar concentration C_A (moles per liter) in our previous discussion.] The continuity equation is usually derived in terms of the mass concentration, because mass is always conserved and moles are not.

The continuity equation for flow in the x direction only is

$$\partial \rho_A/\partial t = -\partial(\rho_A V_{Ax})/\partial x + r_A, \tag{6}$$

where V_{Ax} is the velocity of material A in the x direction in centimeters per second and r_A is the rate of chemical and/or biological mass production or destruction of material A in milligrams per liter per second. The r_A component will also contain any source or sink terms applicable to the model.

Equation (6) states that the rate of change of mass ($\partial \rho_A / \partial t$) within the volume element is equal to the difference between output and input [$-\partial(\rho_A V_{Ax})/\partial x$] plus the mass being produced (r_A), including any sources or sinks.

The continuity equation can be expressed in terms of R_A and the molar concentration C_A as

$$\partial C_A / \partial t = -\partial(C_A V_{Ax})/\partial x + R_A \tag{7}$$

by dividing Eq. (6) by the molecular weight M_A of A and noting that $C_A = \rho_A / M_A$ and that $R_A = r_A / M_A$ is the molar production or destruction of A in the volume element in moles per liter per second (including sources and sinks).

In three dimensions Eqs. (6) and (7) generalize to

$$\frac{\partial \rho_A}{\partial t} = -\left[\frac{\partial}{\partial x}(\rho_A V_{Ax}) + \frac{\partial}{\partial y}(\rho_A V_{Ay}) + \frac{\partial}{\partial z}(\rho_A V_{Az})\right] + r_A \tag{8}$$

and

$$\frac{\partial C_A}{\partial t} = -\left[\frac{\partial}{\partial x}(C_A V_{Ax}) + \frac{\partial}{\partial y}(C_A V_{Ay}) + \frac{\partial}{\partial z}(C_A V_{Az})\right] + R_A. \tag{9}$$

These equations of continuity are for a single material in a multimaterial system.

The continuity equation for all of the material can be found by adding the equations for all the materials in the system to yield

$$\partial \rho / \partial t = -\partial(\rho V_x)/\partial x \tag{10}$$

for the simple case of flow in the x direction, where ρ, the sum of all the mass concentrations, equals $\rho_A + \rho_B + \cdots$ and V_x, the mass average velocity in the x direction, equals $\rho^{-1}\rho_A V_{Ax} + \rho_B V_{Bx} + \cdots$. This velocity is the same as the fluid velocity used in Eq. (3).

The sum of the terms $r_A + r_B + \cdots$ vanishes, because the mass of all the reacting components must be conserved.

In three dimensions Eq. (10) becomes

$$\frac{\partial \rho}{\partial t} = -\left[\frac{\partial}{\partial x}(\rho V_x) + \frac{\partial}{\partial y}(\rho V_y) + \frac{\partial}{\partial z}(\rho V_z)\right]. \tag{11}$$

For the case in which the mass density ρ remains constant, $\partial \rho / \partial t = 0$, Eqs. (10) and (11) become

$$\partial V_x / \partial x = 0, \tag{12}$$

$$\partial V_x / \partial x + \partial V_y / \partial y + \partial V_z / \partial z = 0. \tag{13}$$

Equations (12) and (13) are the continuity equations for constant mass in one and three dimensions, respectively.

The condition in which no changes occur over time in any of the individual materials is called a *steady state*. We shall examine one material in the system and express this mathematically as $\partial \rho_A / \partial t = 0$. Steady-state conditions would reduce Eqs. (6) and (8) to

$$r_A = \frac{\partial}{\partial x}(\rho_A V_{Ax})$$

(14)

and

$$r_A = \frac{\partial}{\partial x}(\rho_A V_{Ax}) + \frac{\partial}{\partial y}(\rho_A V_{Ay}) + \frac{\partial}{\partial z}(\rho_A V_{Az}).$$

(15)

C. Fick's Laws of Diffusion

We shall now develop a more useful form of the continuity equation for a single material A that does not contain the flux term $\rho_A V_{Ax}$. The flux term is eliminated by using the following form of Fick's first law of diffusion:

$$\rho_A V_{Ax} = -\rho \mathcal{D}_{AM} \frac{\partial}{\partial x}\left(\frac{\rho_A}{\rho}\right) + \rho_A V_x,$$

(16)

where $\mathcal{D}_{AM}$ is the mass diffusivity of material A in square centimeters per second. The negative sign indicates that the diffusion of material A relative to the moving fluid is in the direction of decreasing mass concentration ratio.

Substitution of Fick's law as given by Eq. (16) into Eq. (6) and simplifying yields

$$\frac{\partial \rho_A}{\partial t} + V_x \frac{\partial \rho_A}{\partial x} + \rho_A \frac{\partial V_x}{\partial x} = \frac{\partial}{\partial x}\left[\rho \mathcal{D}_{AM} \frac{\partial}{\partial x}\left(\frac{\rho_A}{\rho}\right)\right] + r_A.$$

(17)

For the common case in which the total density ρ and diffusivity $\mathcal{D}_{AM}$ are constant, this equation simplifies to

$$\frac{\partial \rho_A}{\partial t} + V_x \frac{\partial \rho_A}{\partial x} = \mathcal{D}_{AM} \frac{\partial^2 \rho_A}{\partial x^2} + r_A,$$

(18)

where Eq. (12) was used to eliminate the velocity gradient term $\partial V_x / \partial x$.

In three dimensions Eq. (18) generalizes to

$$\frac{\partial \rho_A}{\partial t} + V_x \frac{\partial \rho_A}{\partial x} + V_y \frac{\partial \rho_y}{\partial y} + V_z \frac{\partial \rho_A}{\partial z} = \mathcal{D}_{AM}\left(\frac{\partial^2 \rho_A}{\partial x} + \frac{\partial^2 \rho_A}{\partial y} + \frac{\partial^2 \rho_A}{\partial z}\right) + r_A.$$

(19)

Equations (18) and (19) are written in terms of the molar concentration by dividing by M_A ($C_A = \rho_A/M_A$) as follows:

$$\frac{\partial C_A}{\partial t} + V_x \frac{\partial C_A}{\partial x} = \mathcal{D}_{AM} \frac{\partial C_A}{\partial x} + R_A, \tag{20}$$

$$\frac{\partial C_A}{\partial t} + V_x \frac{\partial C_A}{\partial x} + V_y \frac{\partial C_A}{\partial y} + V_z \frac{\partial C_A}{\partial z}$$
$$= \mathcal{D}_{AM} \left(\frac{\partial^2 C_A}{\partial x^2} + \frac{\partial^2 C_A}{\partial y^2} + \frac{\partial^2 C_A}{\partial z^2} \right) + R_A. \tag{21}$$

These equations can now be used to yield another form of the total time derivative DC_A/Dt previously given by Eq. (3).   Substitution of the left-hand side of Eq. (20) into Eq. (3) yields

$$\frac{DC_A}{Dt} = \mathcal{D}_{AM} \frac{\partial^2 C_A}{\partial x^2} + R_A. \tag{22}$$

Similarly, in three dimensions

$$\frac{DC_A}{Dt} = \mathcal{D}_{AM} \left(\frac{\partial^2 C_A}{\partial x^2} + \frac{\partial^2 C_A}{\partial y^2} + \frac{\partial^2 C_A}{\partial z^2} \right) + R_A. \tag{23}$$

For a quiescent system (in which the velocity of the fluid is equal to zero) with a zero mass production rate R_A, Eqs. (20) and (21) reduce to the following forms of Fick's second law of diffusion:

$$\frac{\partial C_A}{\partial t} = \mathcal{D}_{AM} \frac{\partial^2 C_A}{\partial x^2}, \tag{24}$$

$$\frac{\partial C_A}{\partial t} = \mathcal{D}_{AM} \left(\frac{\partial^2 C_A}{\partial x^2} + \frac{\partial^2 C_A}{\partial y^2} + \frac{\partial^2 C_A}{\partial z^2} \right). \tag{25}$$

The steady-state forms of the continuity equations in terms of the molar concentration are obtained from Eqs. (22) and (23) as

$$V_x \frac{\partial C_A}{\partial x} = \mathcal{D}_{AM} \frac{\partial^2 C_A}{\partial x^2} + R_A \tag{26}$$

and

$$V_x \frac{\partial C_A}{\partial x} + V_y \frac{\partial C_A}{\partial y} + V_z \frac{\partial C_A}{\partial z} = \mathcal{D}_{AM} \left(\frac{\partial^2 C_A}{\partial x^2} + \frac{\partial^2 C_A}{\partial y^2} + \frac{\partial^2 C_A}{\partial z^2} \right) + R_A. \tag{27}$$

Equation (27) represents the general equation that we shall utilize to model the movement of materials in streams, lakes, and engineered reactors.   We shall now see how water quality models are developed and describe the parameters important to their functioning.

III. MODEL DEVELOPMENT

A model of an environmental system is a quantitative representation of that system. Models are required to predict the outcome of various processes operating within a system and the change in concentration of substances within fluid systems.

In this section we shall develop mathematical models to represent certain natural processes. These processes are idealized in some cases because of the lack of quantitative information and also to keep the models workable.

Most models are based on the law of conservation of mass and predict the change in concentration of a substance that occurs within a specific volume. In essence, we observe the mass of a substance within a volume at a specific time, consider the material transport into and out of the volume, and finally take into account the creation or destruction of the substance within the volume due to biological and/or chemical processes.

We shall now develop some of the more common water quality models and show how material inputs and transport are handled. Our first example will be the dissolved oxygen system.

A. The Dissolved Oxygen System

The importance of the dissolved oxygen content of a body of water as an overall index of water quality is well known. Modeling the dissolved oxygen system is a somewhat complex but excellent example of the modeling process.

The time rate of change of oxygen in an element of fluid in a one-dimensional flow moving through a stream reach can be defined as

$$\frac{dC}{dt} = \mathscr{D}_{AM} \frac{\partial^2 C}{\partial x^2} + R_A. \tag{28}$$

For this example the term R_A is considered to be composed of an aeration rate R_1 and a deoxygenation rate R_2. We begin with

$$R_1 = k_2(C_s - C), \tag{29}$$

where C is the concentration of dissolved oxygen in solution at time t in milligrams per liter, C_s the saturation concentration of oxygen in equilibrium with the atmosphere, and k_2 the aeration coefficient, and

$$R_2 = k_1 L, \tag{30}$$

where k_1 is the deoxygenation constant per unit of time and L the first-stage

BOD remaining after time t in milligrams per liter. Equation (28) can now be written as

$$\frac{dC}{dt} = \mathcal{D}_{AM}\frac{\partial^2 C}{\partial x^2} + k_2(C_s - C) - k_1 L. \tag{31}$$

For one-dimensional flow the total time derivative can be written as

$$\frac{dC}{dt} = \frac{\partial C}{\partial x} V_x + \frac{\partial C}{\partial t}, \tag{32}$$

and for the many applications in which the volumetric flow rate Q and cross-sectional area of the stream A remain constant, Eq. (31) can be simplified to the form

$$V\frac{\partial C}{\partial x} + \frac{\partial C}{\partial t} = \mathcal{D}_{AM}\frac{\partial^2 C}{\partial x^2} + k_2(C_s - C) - k_1 L, \tag{33}$$

where V is the constant velocity Q/A.

We shall now consider two important physical states that substantially reduce the complexity of the model given by Eq. (33). First, for the treatment of bodies of water such as streams, in addition to satisfying the previous assumptions concerning a constant flow rate Q and a constant cross-sectional area A, the longitudinal diffusion term from Eq. (33), $\partial^2 C/\partial x^2$, is considered negligible. Second, for those systems in which the input and output are constant, there exists a steady-state condition ($\partial C/\partial t = 0$), and Eq. (33) becomes

$$V\frac{\partial C}{\partial x} = k_2(C_s - C) - k_1 L. \tag{34}$$

It is convenient to use a term for the oxygen deficit,

$$D = C_s - C, \tag{35}$$

because it allows for a comparison of the concentration C with the oxygen saturation value at any time t. The investigator will now be better able to isolate the dissolved oxygen dependence on the variable x. Substituting Eq. (35) into Eq. (34) and rearranging terms gives

$$(\partial D/\partial x) + (k_2/V)D = (k_1/V)L. \tag{36}$$

Equation (36) models the oxygen deficit along a stream reach at any time t.

The integration of Eq. (36) should be done in two parts, the integration of the homogeneous equation

$$\partial D/\partial x + (k_2/V)D = 0 \tag{37}$$

and any particular integral of the equation in this form. Equation (37) is a first-order, first-degree linear differential equation and can be easily integrated as

$$D = A \exp(-k_2 x/V), \tag{38}$$

where A is an arbitrary constant to be used in satisfying the boundary condition. One way to find a particular integral of Eq. (36) is to first determine an equation for L. This can be developed similarly to Eq. (34) for C as

$$V(\partial L/\partial x) = -(k_1 + k_3)L + L_a, \tag{39}$$

where k_3 is the BOD removal rate constant per unit of time owing to the sedimentation of solids and L_a the BOD addition rate to the reach from nonpoint sources. This equation can be written in a form more amenable to obtaining an integral as

$$\partial L/\partial x + (k_1 + k_3)/VL = (1/V)L_a. \tag{40}$$

The assumption that the terms other than L in Eq. (39) are constant over a given reach leads to the solution

$$L(x) = \left(L_0 - \frac{L_a}{k_1 + k_3} \right) \exp\left[-\frac{(k_1 + k_3)}{V}x \right] + \frac{L_a}{k_1 + k_3}, \tag{41}$$

where L_0 is the notation for the boundary value $L(0)$ of $L(x)$ at $x = 0$. We can now rewrite Eq. (36), after substituting the expression for $L(x)$ given by Eq. (41), as

$$\frac{\partial D}{\partial x} + \frac{k_2}{V}D = \frac{k_1}{V}\left(L_0 - \frac{L_a}{k_1 + k_3} \right)$$
$$\times \exp\left[-\frac{(k_1 + k_3)}{V}x \right] + \frac{k_1 L_a}{V(k_1 + k_3)}. \tag{42}$$

The cumbersome nature of Eq. (42) can be eliminated by defining some new terms:

$$\bar{L}_a = (k_1/V)[1/(k_1 + k_3)]L_a, \qquad \bar{k} = (k_1 + k_3)/V,$$

$$\bar{k}_2 = k_2/V, \qquad \bar{L}_0 = k_1 L_0/V, \qquad \bar{L} = \bar{L}_0 - \bar{L}_a. \tag{43}$$

We now have

$$\partial D/\partial x + \bar{k}_2 D = \bar{L} \exp(-\bar{k}x) + \bar{L}_a. \tag{44}$$

The following particular integral of Eq. (44) can be found:

$$D(x) = \bar{k}/(k_2 - k) \exp(-\bar{k}x) + \bar{L}_a/k_2. \tag{45}$$

Adding this particular integral to the homogeneous integral [Eq. (38)], applying the boundary condition that $D(0) = D_0$ at $x = 0$, and simplifying terms yields

$$D(x) = D_0 \exp(-\bar{k}_2 x) - \left(\frac{\bar{L}}{\bar{k}_2 - \bar{k}} + \frac{\bar{L}_a}{\bar{k}_2} \right)$$

$$\times \exp(-\bar{k}_2 x) + \frac{\bar{L}}{\bar{k}_2 - \bar{k}} \exp(-\bar{k} x) + \frac{\bar{L}_a}{\bar{k}_2}. \tag{46}$$

Equation (46) can be rewritten in a form using only the originally defined terms as follows:

$$D(x) = D_0 \exp\left(\frac{-k_2 x}{V} \right) + \frac{k_1}{k_2 - (k_1 + k_3)} \left(L_0 - \frac{L_a}{k_1 + k_3} \right)$$

$$\times \left(\exp\left[-\frac{(k_1 + k_3)x}{V} \right] - \exp\left[-\left(\frac{k_2 x}{V} \right) \right] \right)$$

$$+ \frac{k_1 L_a}{k_2(k_1 + k_3)} \left\{ 1 - \exp\left[-\frac{(k_2 x)}{V} \right] \right\} \bigg). \tag{47}$$

Once the aeration, deoxygenation, and sedimentation rates are determined in the field, a plot of oxygen deficit versus distance of stream reach can be developed. The first derivative of this expression will yield the maximum deficit observed.

It can be seen that the complete solution of the first-order, linear differential equation (36) is given by

$$D(x) = \exp(-k_2 x/V)[k_1/V \int L \exp(k_2 x/V) \, dx + A], \tag{48}$$

where A is an arbitrary constant. Note that for $L = 0$ this solution reduces to that given by Eq. (38). It can easily be shown that substitution of the form for L given in Eq. (41) leads directly to the solution given by Eq. (47).

Certain physical situations may result in higher-than-first-order dependence on L. Consider, for example, the right-hand side of Eq. (36) having the form $(k_n/V)L^n$ instead of $(k_1/V)L$, where k_n is the appropriate rate constant. Because the right-hand side of Eq. (36) is still a function of x, it would only be necessary to replace $k_1 L/V$ by $(k_n/V)L^n$ in Eq. (42) to determine the new solution for $D(x)$. Assume L is of a form similar to that given by Eq. (41), that is,

$$L = a \exp(-bx) + c, \tag{49}$$

where a, b, and c are constants. General integration procedures can be developed for the nth degree case, but the terms that would have to be integrated would be of the form

$$L^n = a^n \exp(-nbx) + na^{n-1}c \exp[-(n-1)bx] + n(n-1)/2$$
$$+ a^{n-2}c^2 \exp[-(n-2)bx] + \cdots . \qquad (50)$$

The integration in Eq. (48) of terms such as those in Eq. (50) is basically the same as that for terms in the first-degree equation (49).

Thus it is seen that second- and higher-order dependence on L in the oxygen deficit equation can be easily treated. This is important to consider as more representative process relationships are developed. Often models assume zero- or first-order reaction kinetics for simplicity when more representative functions can be accommodated.

B. Inputs to and Designation of a Stream Reach

The basis for the choice of certain stream reaches, with regard to simplifying the physical model, has already been discussed. Additional considerations, for example, have to do with the location of waste discharge points along the stream. That is, a simplification in the modeling process can be achieved if a new reach begins at the location of each of the waste discharge points. The waste discharge is then treated as a boundary condition. Cases in which the discharge is time dependent will be discussed in succeeding sections. Other situations that would suggest starting a new reach include abrupt changes in flow, such as that caused by a tributary, and abrupt changes in area (see Fig. 9.1).

Gradual changes in conditions along a reach, such as stormwater runoff, can be treated by the addition of terms in the model that deal with a

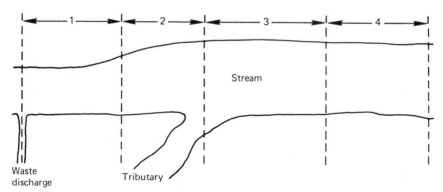

FIGURE 9.1   SELECTING BOUNDARY CONDITIONS FOR A STREAM REACH.

variation in the flow rate. This is usually a nonpoint source input, a case we shall now consider.

Surface water runoff is a distributed source that is associated with a change in the flow of the reach over time. Recalling the basic continuity equation (6),

$$\partial \rho_A / \partial t = -\partial(\rho_A V_{Ax})/\partial x + r_A,$$

it can be seen that a term such as

$$\hat{\rho}_A \left(\frac{\partial Q}{\partial x} \Delta x \right) \Delta t, \tag{51}$$

where $\hat{\rho}_A$ is the amount of material [ρ_A is added to a Δx slice of the reach by the increased flow $(\partial Q/\partial x) \Delta x$ in time Δt], represents a distributed source. Also, the term

$$\tilde{\rho}_A \Delta x \Delta t, \tag{52}$$

where $\tilde{\rho}_A$ is the rate of material produced or destroyed in a Δx slice of the reach, represents a distributed source or sink within the reach but not associated with an increase in flow. Benthic deposits distributed along a reach, for instance, represent a physical situation that can be modeled by this term.

Equation (6) can now be rewritten in a form that isolates the two distributed source terms as follows:

$$\frac{\partial \rho_A}{\partial t} = -\frac{\partial}{\partial x}(\rho_A V_{Ax}) + \hat{\rho}_A \frac{\partial Q}{\partial x}\left(\frac{1}{A}\right) + \tilde{\rho}_A + \tilde{r}_A, \tag{53}$$

where $\tilde{r}_A$ is equal to r_A minus the two distributed source terms. Fick's law [Eq. (16)] can again be used to eliminate the flux term $\rho_A V_{Ax}$, yielding

$$\frac{\partial \rho_A}{\partial t} + \rho_A \frac{\partial V_x}{\partial x} + V_x \frac{\partial \rho_n}{\partial x} = \frac{\partial}{\partial x}\left[\rho \mathcal{D}_{AM} \frac{\partial}{\partial x}\left(\frac{\rho_n}{\rho}\right)\right]$$
$$+ \hat{\rho}_A \frac{\partial Q}{\partial x}\left(\frac{1}{A}\right) + \tilde{\rho}_A + \tilde{r}_A. \tag{54}$$

For constant density ρ and diffusivity $\mathcal{D}_{AM}$, Eq. (54) reduces to

$$\frac{\partial \rho_A}{\partial t} + V_x \frac{\partial \rho_A}{\partial x} = \mathcal{D}_{AM} \frac{\partial^2 \rho_A}{\partial x} + \frac{\hat{\rho}_A}{A} \frac{\partial Q}{\partial x} + \tilde{\rho}_A + \tilde{r}_A. \tag{55}$$

For a nondispersive system, Eq. (54) can be written as

$$\frac{\partial \rho_A}{\partial t} = -\frac{\partial(\rho_A V_x)}{\partial x} + \frac{\partial_A}{A} \frac{\partial Q}{\partial x} + \tilde{\rho}_A + \tilde{r}_A. \tag{56}$$

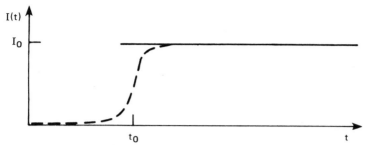

FIGURE 9.2   SCHEMATIC DIAGRAM OF A CONSTANT WASTE DISCHARGE INPUT: — —, REALISTIC PHYSICAL SITUATION; ——, MATHEMATICAL IDEALIZATION OF THE PROCESS.

For reaches or systems of constant crossectional area A,

$$\frac{\partial \rho_A}{\partial t} = -\frac{1}{A}\frac{\partial(Q\rho_A)}{\partial x} + \frac{1}{A}\hat{\rho}_A\frac{\partial Q}{\partial x} + \tilde{\rho}_A + \tilde{r}_A. \tag{57}$$

Recall that the origin of each system usually coincides with a waste discharge point. Therefore, the appropriate boundary condition would be

$$\rho_A(x, t) = \rho_{A0}(t), \qquad x = 0.$$

The term $\rho_{A0}(t)$ represents the concentration of ρ_A in the stream at the waste discharge point $x = 0$.

1. Step Inputs

Previous discussions have assumed that the waste discharge is constant after some arbitrary starting time t_0. This type of input is illustrated in Fig. 9.2. In equation form this type of input can be written

$$I(t) = I_0 U(t - t_0), \tag{58}$$

where $U(t - t_0)$ is the unit step function

$$U(t - t_0) = \begin{cases} 0 & \text{for} \quad t < t_0, \\ 1 & \text{for} \quad t > t_0. \end{cases} \tag{59}$$

The function $U(t - t_0)$ is presented graphically in Fig. 9.3.

2. Rectangular Pulse Inputs

The unit step function can also be used to represent a rectangular pulse input, as shown in Fig. 9.4. In equation form this input is

$$I(t) = I_0[U(t - t_0) - U(t - t_1)], \tag{60}$$

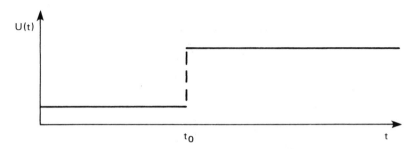

FIGURE 9.3   UNIT STEP FUNCTION FOR FLOW IN A STREAM REACH ASSUM-
ING A CONSTANT WASTE DISCHARGE AFTER AN ARBITRARY STARTING
TIME t_0.

or

$$I(t) = \begin{cases} 0 & \text{for } t < t_0, \\ I_0 & \text{for } t_0 < t < t_1, \\ 0 & \text{for } t > t_1, \end{cases} \qquad (61)$$

and represents a constant input over a specified time interval $\Delta t = t_1 - t_0$.
Consider the case in which $I(t)$ is defined by

$$I(t) = \begin{cases} 0 & \text{for } t < t_0, \\ I_0/\Delta t & \text{for } t_0 < t < t_1, \\ 0 & \text{for } t > t_1. \end{cases} \qquad (62)$$

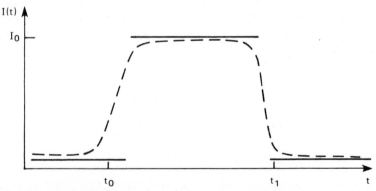

FIGURE 9.4   RECTANGULAR PULSE FUNCTION FOR FLOW IN A STREAM
REACH ASSUMING A CONSTANT WASTE DISCHARGE OVER A SPECIFICIED
TIME INTERVAL FROM t_0 TO t_1.

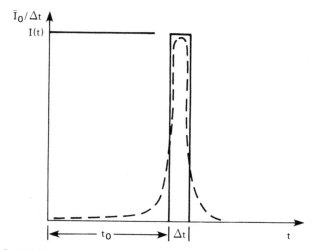

FIGURE 9.5   RECTANGULAR PULSE FUNCTION FOR FLOW IN A STREAM REACH FOR A SMALL TIME INTERVAL Δt.

This representation, in which Δt is small, is illustrated in Fig. 9.5. For very small intervals of $\Delta t = t_1 - t_0$, greater and greater concentrations of discharge are introduced, but the total amount always remains the same: $(\bar{I}_0/\Delta t) \Delta t = \bar{I}_0$.

3. Pulse Inputs

It is sometimes convenient to pursue the idea of an increasing waste input intensity over a decreasing time interval to the limit, keeping the product of the intensity and duration constant. The mathematical function used to describe this idealized input is the Dirac delta function $\delta(t - t_0)$, written symbolically as

$$\delta(t - t_0) = \begin{cases} 0 & \text{for} \quad t \neq t_0, \\ \infty & \text{for} \quad t = t_0, \end{cases} \tag{63}$$

with the integral of $\delta(t - t_0)$ normalized to unity:

$$\int_{-\infty}^{\infty} \delta(t - t_0)\, dt = 1. \tag{64}$$

This last property suggests the alternate labeling of $\delta(t - t_0)$ as a unit impulse function.

An important property of this function is its ability to isolate a particular

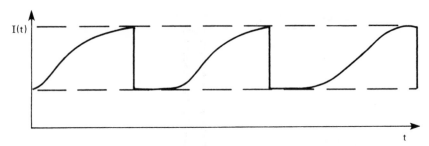

FIGURE 9.6   PERIODIC FUNCTION FOR FLOW IN A STREAM REACH IN WHICH
THE INPUT CHARACTERISTICS ARE REPEATABLE IN TIME.

value of a function $I(t)$ when integrated,

$$\int_{-\infty}^{\infty} I(t)\, \delta(t - t_0)\, dt = I(t_0), \tag{65}$$

which is sometimes called the sifting property.

4. Periodic Inputs

An example of an input whose characteristics are repeatable in time is shown in Fig. 9.6. Inputs of this type are usually best treated through the representation of $I(t)$ by the trigonometric series

$$\frac{1}{2} a_0 + \sum_{n=1}^{\infty} \left(a_n \cos \frac{2\pi n t}{T} + b_n \sin \frac{2\pi n t}{T} \right). \tag{66}$$

This series becomes the Fourier series of $I(t)$ if the coefficients a_0, a_n, and b_n are evaluated by the formulas

$$a_n = \left(\frac{2}{T}\right) \int_{-T/2}^{T/2} I(t) \cos \frac{2\pi n t}{T}\, dt \qquad n \geq 0, \tag{67}$$

$$b_n = \left(\frac{2}{T}\right) \int_{-T/2}^{T/2} I(t) \sin \frac{2\pi n t}{T}\, dt \qquad n > 0. \tag{68}$$

In these expressions T is the period of the function $I(t)$.

IV. CASE STUDY: THE PHOSPHORUS MODEL

The previous sections have defined the approaches common to water quality model development. Although many situations exist in which the basic model must be altered, either due to unique waste inputs or hydrological situations, the basic approach developed in this chapter can be followed in all cases. As an example, consider the development of a stream model that reflects the movement of phosphorus. Such models are important for managing bodies of water with respect to the eutrophication process.

Consider the model defining the transport of phosphorus in meandering streams as developed by Rosendahl and Waite (1978). In the expanded form with all reaction, sink, and source terms, the equation (assuming a steady state) is

$$k_{c_1} V_x \frac{dP}{dx} = k_{c_2} \mathcal{D}_L \frac{d^2P}{dx^2} - k_1 P^3 - \frac{k_2 P}{k_P + P} k_{(\epsilon)}, \qquad (69)$$

where P is the concentration of phosphorus, V_x the average river velocity in the downstream direction, $\mathcal{D}_L$ the longitudinal dispersion coefficient, k_1 the rate constant for phosphorus precipitation, k_2 the maximum rate of phosphorus uptake by algae, $k_{(\epsilon)}$ the sum of the zero-order rate processes that affect phosphorus movement, k_P the Michaelis–Menton uptake constant, and k_{c_1} and k_{c_2} arbitrary constants used for model calibration. Tributary sources of phosphorus will be considered to be point sources and will be accounted for in the boundary conditions of the stream reach.

It is instructive to solve a linearized version of Eq. (69). The equation in linear form will be directly solvable by the behavior of the relationship between phosphorus and downstream distance.

Equation (69) can be linearized by simply taking the Taylor series expansion of the two nonlinear terms, that is, the third-power chemical precipation term, $k_1 P^3$, and the Michaelis–Menton term for phosphorus uptake by algae, $(k_2 P / k_P + P)$. Expanding any appropriate function of P in a Taylor series about any convenient value of P, say, P_0 (the value of P at $x = 0$), and including only the first-order term gives

$$f(P) = F(P_0) + (P - P_0) f'(P_0) + \text{higher-order terms.}$$

Therefore,

$$p^3 = P_0^3 + 3(P - P_0)P_0^2$$

and

$$P/(k_P + P) = P_0/(k_P + P_0) + k_P(P - P_0)/(k_P + P_0)^2.$$

Simplifying algebraically and substituting back into Eq. (69) yields

$$k_{c_1} V_x \frac{dP}{dx} = k_{c_2} \mathcal{D}_L \frac{d^2D}{dx^2} - k_1(3P_0^2 P - 2P_0^3) - k_2 \frac{P_0^2 + k_P P}{(k_P + P_0)^2} + k_{(\epsilon)} = 0. \quad (70)$$

Equation (70) can now be written as

$$A \frac{d^2P}{dx^2} + B \frac{dP}{dx} + CP = D, \tag{71}$$

where A, B, C, and D are the following constants:

$$A = k_{c_2}\mathcal{D}_L, \qquad B = k_{c_1}V_x,$$

$$C = -k_1 3P_0^2 - \frac{k_2 k_P}{(k_P + P_0)^2}, \qquad D = -2k_1 P_0^3 + k_2\frac{P_0^2}{(k_P + P_0)^2} - k_{(\epsilon)}.$$

It can be seen that Eq. (71) is a second-order, first-degree, linear, constant-coefficient ordinary differential equation.

The particular solution, that is, any function of P that satisfies the right-hand side of the equation, is $P = D/C$. The solution of the homogeneous equation [the left-hand side of Eq. (71)] is

$$P_2 C_1 \exp(m_1 x) + C_2 \exp(m_2 x),$$

where

$$m_1 = \frac{-B + \sqrt{B^2 - 4AC}}{2A}, \qquad m_2 = \frac{-B - \sqrt{B^2 - 4AC}}{2A}.$$

Thus, the general solution is

$$P = C_1 \exp(m_1 x) + C_2 \exp(m_2 x) + D/C. \tag{72}$$

The values of C_1 and C_2 are determined from the boundary conditions $P(0) = P_0$ and $P'(0) = P_0'$, where $P'(x) = dP/dx$. Using these boundary conditions in Eq. (72) yields

$$P_0 = C_1 + C_2 + D/C, \qquad P_0' = m_1 C_1 + m_2 C_2.$$

Solving for C_1 and C_2 yields

$$C_1 = \frac{m_2[(D/C) - P_0] + P_0'}{m_1 - m_2}, \qquad C_2 = \frac{m_1[P_0 - (D/C)] - P_0'}{m_1 - m_2}.$$

Plugging in the values from our phosphorus model, the parameters for the general solution are defined as follows:

$$m_1 = \frac{k_{c_1}V_x + \sqrt{k_{c_1}^2 V_x^2 + 4k_{c_2}\mathcal{D}_L(3k_1 P_0^2) + k_2 k_P(k_P + P_0)^{-2}}}{2k_{c_2}},$$

$$m_2 = \frac{k_{c_1}V_x - \sqrt{k_{c_1}^2 V_x^2 + 4k_{c_2}\mathcal{D}_L(3k_1 P_0^2) + k_2 k_P(k_P + P_0)^{-2}}}{2k_{c_2}\mathcal{D}_L},$$

$$m_1 = m_2 = \frac{\sqrt{k_{c_1}^2 V_x + 4k_{c_2}\mathcal{D}_L(3k_1 P_0^2) + k_2 k_P(k_P + P_0)^{-2}}}{k_{c_2}\mathcal{D}_L},$$

$$D/C = \frac{-2k_1 P_0^3 + k_2 P_0^2(k_P + P_0)^{-2} - k\delta}{-3k_1 P_0^2 - k_2 k_P(k_P + P_0)^{-2}},$$

$$C_1 = m_2[(D/C) - P_0] + P_0'(m_1 - m_2)^{-1},$$

$$C_2 = m_1[P_0 - (D/C)] - P_0'(m_1 - m_2)^{-1}.$$

V. CASE STUDY: LONGITUDINAL DISPERSION

Recall the one-dimensional continuity equation for the case of constant total density, with diffusivity $\mathcal{D}_{AM}$ given by Eq. (28) with $R_A = 0$:

$$\frac{\partial C_A}{\partial t} + V_x \frac{\partial C_A}{\partial x} = \mathcal{D}_{AM} \frac{\partial^2 C_A}{\partial x^2}.$$

This equation describes the variation of C with time and with the "longitudinal" coordinate x in the direction of flow. This equation can be made non-dimensional by introducing the following variables:

$$C = C_A / C_{A_0}, \qquad \tau = t/\bar{t}, \qquad \bar{t} = L/V_x,$$

$$\xi = x/L, \qquad P_e = (\mathcal{D}_{AM}/V_x L)^{-1},$$

where C_A is the concentration at time t, C_{A_0} the initial concentration at time $t = 0$, t the detention time, L the characteristic length, V_x the velocity in x direction, and P_e the Pectlet number, a dispersion characteristic of the system.
Substitution yields

$$\frac{\partial C}{\partial \tau} + \frac{\partial C}{\partial \xi} = \frac{1}{P_e} \frac{\partial^2 C}{\partial \xi^2}. \tag{73}$$

The extreme values of the Pectlet number, $P_e \to \infty$ (no axial dispersion) and $P_e \to 0$ (complete dispersion), then represent ideal plug flow and a completely mixed system, respectively.

This section is primarily concerned with estimating values of the Pectlet number for a fluid system that has small but finite dispersion by using a solution to Eq. (73). The solution assumes that a conservative tracer is injected into a stream and followed with time.

The instantaneous injection of this tracer is mathematically described by the pulse input. The solution to Eq. (73) for small dispersion (large values

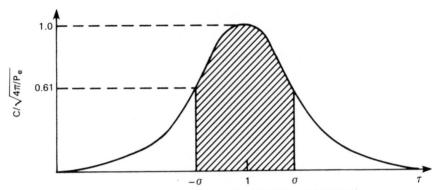

FIGURE 9.7   PROBABILITY DENSITY FUNCTION FOR LONGITUDINAL DISPERSION OF A SUBSTANCE OF CONCENTRATION C IN A STREAM REACH.

of P_e) owing to a pulse input can be shown to be

$$C = \frac{1}{\sigma\sqrt{2\pi}} e^{-1/2}[(\tau - m)/\sigma]^2. \tag{74}$$

The solution can easily be verified by substitution back into Eq. (73), where $m = \xi$ and $\sigma^2 = 2/P_e$. Equation (74) represents a family of Gaussian (or normal) probability density functions with mean m and variance σ^2. If values of C at $\xi = 1$ ($x = L$) are plotted versus time τ, a curve such as that shown in Fig. 9.7 is obtained. It is seen that the variance, $\sigma^2 = 2/P_e$,

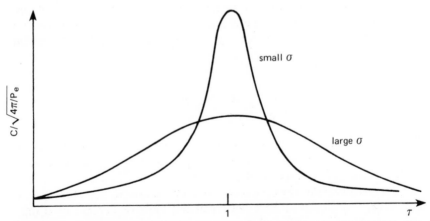

FIGURE 9.8   PROBABILITY DENSITY FUNCTIONS FOR LONGITUDINAL DISPERSION ILLUSTRATING THE EFFECT OF CHANGES IN THE PECTLET NUMBER P_e. $\sigma^2 = 2/P_e$.

completely describes the curve. The larger the value of P_e, the more a curve has its area concentrated about the mean (see Fig. 9.8). The values of C always approach zero asymptotically for $\tau - 1 \rightarrow \pm \infty$ and are symmetric about the mean $\tau = 1$. For practical applications, however, $\tau > 0$, and values of C are negligible at $\tau = 0$ for large values of P_e. The curves are all concave downward for τ within one standard deviation $(\pm \sigma)$ of the mean and concave upward outside of these values. The total area under the curve is 1. The area under the curve between the points $\tau = \pm \sigma$ of the mean is always 68% of the total area, and the value of C is always 61% of its maximum at these points.

Solving for the Pectlet number P_e yields the expression $P_e = 2/\sigma^2$. Thus, the Pectlet number can be determined by plotting time versus concentration for a pulse input and calculating values from the properties of the C curve.

REFERENCES

Rosendahl, P. C., and Waite, T. D. (1978). Transport characteristics of phosphorus in channelized and meandering streams. *Water Res. Bull.* **14**(5).

Chapter 10

Water Quality Standards
and Management Approaches

I. INTRODUCTION

Although the previous chapters have discussed water quality problems and public health concerns, we have not yet discussed the allowable limits of pollutants in water systems. The limits set on various contaminants depend on the purpose for which the water supply is intended. This chapter will address the decision-making processes involved in setting water quality criteria and control measures for maintaining good water quality.

When water quality is to be managed by controlling waste inputs, some bases for the degree of treatment required must be developed. Sound engineering practice dictates that the treatment of waste be both as economic as possible and effective for protecting the quality of receiving waters. For this to be a reality, the limits for pollutants in receiving waters must be carefully defined and given a sound scientific basis. The basis of these limits should be, first, the public health of the community, and second, the environmental health of biological systems within the receiving water. The limits that are set on water contaminants may have different names.

McKee and Wolfe (1963), in their classic work "Water Quality Criteria," differentiate between water quality standards and water quality criteria as parameters for limiting pollutants in water. They note that criteria refer to limits of pollutants that are not established by legal authority. They contend that "standards" should be the legal term for defining such limits. In any case, it is a "water quality standard" that defines the level of pollutants allowed in a water system.

As mentioned previously, standards are important for deciding on the degree of treatment of polluted inputs required in a particular drainage system. One of the goals of setting standards, for example, is to ensure that water supplies can be effectively treated by common engineering procedures to attain desired use levels. This basic assumption is fundamental to all water quality standards.

The philosophical value of standards was evaluated by McGauhey (1968), who lists certain recognizable virtues of water quality standards:

(1)   they encourage the measurement of quality factors;
(2)   they permit self-control by discharges;
(3)   they preserve fairness in the application of police power;
(4)   they furnish a historical documented story of an event and therefore assist in controlling the future;
(5)   they make possible the definition of a problem;
(6)   they establish goals for system designs;
(7)   they force people to face their own ignorance or, when facing arbitrary standards, to define why they are inappropriate;
(8)   they assess the benefits we receive from a system.

These properties of water quality standards help to satisfy social demands for pure water. Originally, only public health concerns, that is, the prevention of specific illnesses, were important for determining water quality. However, in recent years there have been demands for clean or pure water for all uses of a water resource. There is therefore a strong social pressure to establish standards to maintain clean water throughout water systems.

There are four classes of standards used to achieve the goal of overall water purity:

(1)   standards directly on the disease-causing agents;
(2)   index standards, for cases in which some associated factor is easier to measure than the agent itself (e.g., coliforms);
(3)   standards on precursors or elements that enter into reactions that affect water quality (e.g., BOD and nutrients);
(4)   standards on major environmental factors that produce effects worthy of concern (e.g., solar energy).

Perhaps the most important concern is the attainability of the standard. Although a given level of a pollutant in water can be estimated using medical evidence, the actual safe level cannot always be precisely defined. Therefore, appropriate safety factors are generally added to prevent public health problems. However, the final value used as a standard must be reasonable, and any discharger should be able to achieve the standard through common engineering practices. If extremely stringent standards are established that cannot be obtained by a discharger, their validity will inevitably be challenged. Therefore, unless standards are established using sound scientific and/or medical evidence, they will not withstand legal challenges. We will see that standards for certain water uses, for example, drinking water, can be more easily documented using medical studies than can standards set for recreation.

Assuming that standards are used to control water quality in both natural and man-made systems, two types of standards can be established. These are defined as "stream standards" and "effluent standards." Stream standards refer to water quality standards that are established for receiving waters, whereas effluent standards refer to the quality of the effluents that are discharged to receiving waters. The purpose of both effluent and stream standards is to protect or establish the desired quality of a water system. Stream standards take into account the dilution capacities of streams as well as recognizing that the use of streams varies depending on the requirements of society. Also, stream standards take into account interactions among dischargers for maintaining desired stream quality levels. Effluent standards, on the other hand, are usually more stringent and restrict the type or amount of pollutants that can be discharged. Some effluent standards also establish degrees of treatment required before discharge. However, these standards do not allow for discharger interaction, because each waste stream is regulated separately.

There are advantages and disadvantages to each type of standard, and perhaps the optimal system would be a combination of both effluent and stream standards. Consider a situation in which certain industries are discharging pollutants to a stream, resulting in the degradation of water quality. It is possible that several industries could be discharging the same pollutant, but that treatment for the removal of the pollutant is more difficult in some situations because of the particular industrial process. In this case requiring each industry to meet a given effluent standard would unjustly punish certain dischargers with higher treatment costs and perhaps drive them out of business. On the other hand, some industries may be able to remove the pollutant effectively by altering their industrial process or through simple engineering treatment schemes. If a stream standard were in effect, the quality of the stream with respect to a particular pollutant could be maintained by having one industry treat its effluent to a greater degree than another industry. It might also be more cost-effective for a specific industry to help another industry treat their effluent than it would be if the industry treated its own effluent. The goal would be for the stream water quality to meet the desired standard for the desired use of the stream. If rigid effluent standards were imposed on each industry, however, this type of interaction could not occur.

The obvious drawback to stream standards is that, when the quality criteria for a stream are exceeded and the stream is degraded, it is difficult to discern the actual cause. Sampling the water quality within the stream itself does not determine which discharger is responsible. It is obvious that the policing of established standards is a much easier process when effluent standards are used; it is a simple procedure to analyze various waste streams and determine if certain pollutants are being discharged.

In most cases both effluent and stream standards are used, that is, an administrative agency will regulate pollutant levels in effluents as well as setting basic stream standards, neither of which can be violated. In general, effluent standards reflect upper limits on pollutants that are known to be harmful to aquatic systems, and stream standards reflect levels of physical and biological parameters that are known to be compatible with the desired use of a stream.

Table 10.1 shows typical in-stream water quality standards. This particular list of standards, for the State of Illinois, sets limits for various pollutants as well as for the physical parameters of the stream. Because of the dynamic nature of society, the list of pollutants and their allowable concentration must be updated regularly. As an example, the listing for pesticides may change on a yearly basis as their toxicities are observed in the environment. In addition, there are many regulatory agencies responsible for stream water quality in the United States. These agencies may have jurisdiction over part of a state, an entire state, or a group of states, and all may have their own water quality standards. Table 10.1 shows that the water quality standards for the State of Illinois deal with different water uses as well as different water systems; standards for other states may differ in both specific content and general groupings.

Effluent standards, as mentioned previously, are more specific with respect to pollutant concentration. They may also include the degree of treatment required before an effluent can be discharged to the receiving water system. Table 10.2 shows typical effluent standards for several common pollutants. These standards are usually modified to reflect the water situation at different times of the year, that is, the standards will

TABLE 10.1 IN-STREAM WATER QUALITY STANDARDS FOR THE STATE OF ILLINOIS[a]

PARAMETER	GENERAL	PUBLIC AND FOOD PROCESSING WATER SUPPLIES	SECONDARY CONTACT AND INDIGENOUS AQUATIC LIFE	LAKE MICHIGAN	UNDERGROUND WATERS
Biological					
Fecal coliforms per 100 ml	200	200	1000	20	200
Chemical					
Ammonia nitrogen (as N)	1.5	1.5		0.02	1.5
Arsenic (total)	1.0	0.1	0.25	0.1	0.1
Barium (total)	5.0	1.0	2.0	1.0	1.0
Boron (total)	1.0	1.0		1.0	1.0
Cadmium (total)	0.05	0.010	0.15	0.010	0.010
Carbon chloroform extract		0.7		0.7	0.7
Chloride	500.0	250		12.0	250
Chromium		1.05		0.05	0.05
Chromium (total hexavalent)	0.05	0.05	0.3	0.05	0.05
Chromium (total trivalent)	1.0	1.0	1.0	1.0	1.0
Copper (total)	0.02	0.02	1.0	0.02	0.02
Cyanide	0.025	0.025	0.025	0.025	0.025
Dissolved oxygen ratio	6/5	6/5	3/2	—	6/5
Fluoride	1.4	1.4		1.4	1.4
Fluoride (total)			15.0		
Foaming agents		0.5		0.5	0.5
Iron (total)	1.0	0.3	2.0	0.3	0.3
Iron (dissolved)			0.5		
Lead (total)	0.1	0.05	0.1	0.05	0.05
Manganese (total)	1.0	0.05	1.0	0.05	0.05
Mercury (total)	0.0005	0.0005	0.0005	0.0005	0.0005
Nickel (total)	1.0	1.0	1.0	1.0	1.0
Nitrate nitrogen		10		10.0	10.0

	Col 1	Col 2	Col 3	Col 4	Col 5
Nitrite nitrogen		1		1.0	1.0
Oil (hexane solubles) or equivalent)		0.1	15.0	0.1	0.1
Phenols	0.1	0.001	0.3	0.001	0.001
Phosphorus (as P)	0.05	0.05		0.007	0.05
Selenium (total)	1.0	0.01	1.0	0.01	0.01
Silver			0.1		
Silver (total)	0.005	0.005		0.005	0.005
Sulfates	500.0	250		24.0	250
Zinc	1.0	1.0		1.0	1.0
Zinc (total)			1.0		
Pesticides					
Aldrin		0.001		0.001	0.001
Chlordane		0.003		0.003	0.003
DDT		0.05		0.05	0.05
Dieldrin		0.001		0.001	0.001
Endrin		0.0005		0.0005	0.0005
Heptachlor		0.0001		0.0001	0.0001
Heptachlor epoxide		0.0001		0.0001	0.0001
Lindone		0.005		0.005	0.005
Methoxychlor		0.1		0.1	0.1
Parathion		0.1		0.1	0.1
Toxaphene		0.005		0.005	0.005
Physical					
pH	6.5–9.0	6.5–9.0	6.0–9.0	7.0–9.0	6.5–9.0
Temperature (°F)			93/100		
Total dissolved solids	1000.0	500	3500	180.0	500
Total suspended solids			15.0		
Radioactivity					
Gross beta (pCi l^{-1})	100	100		100	100
Radium 226 (pCi l^{-1})	1	1		1	1
Strontium 90 (pCi l^{-1})	2	2		2	2

[a] All units are expressed in milligrams per liter unless otherwise noted.

239

TABLE 10.2   TYPICAL EFFLUENT STANDARDS
FOR COMMON POLLUTANTS

POLLUTANT	MAXIMUM ALLOWABLE CONCENTRATION (mg/l)
Bacteria	400/100 ml
Ammonia nitrogen (as N)	3.0
Phosphorus (as P)	1.0
Arsenic (total)	0.25
Barium (total)	2.0
Cadmium (total)	0.15
Chromium (total hexavalent)	0.3
Chromium (total trivalent)	1.0
Copper (total)	1.0
Cyanide	0.025
Fluoride (total)	15.0
Iron (total)	2.0
Iron (dissolved)	0.5
Lead (total)	0.1
Manganese (total)	1.0
Mercury (total)	0.0005
Nickel (total)	1.0
Oil (hexane solubles or equivalent)	15.0
pH	5 – 10
Phenols	0.3
Selenium (total)	1.0
Silver	0.1
Zinc (total)	1.0
Total suspended solids	15.0

change with the season and the population.   It should be noted that standards are written on maximum allowable concentrations and do not directly reflect on the total load of a particular pollutant.

Because there is substantial variability in both effluent and stream standards in the United States, we show in Tables 10.3a and 10.3b a comparison of the requirements for water and effluent quality in 10 selected states.   The variations in the standards evolved because of the different climates of the states, the intended uses of the water resources, and the types of water supplies available.   Table 10.3 shows in particular that not only do allowable limits vary between states but also that some states do not regulate certain pollutants that are regulated by other states.   Again, this reflects each state's desire to maintain water quality as well as its degree of industrialization. Generally, as states become more developed, with larger populations, more stringent water quality standards are issued. Those states that are heavily industrialized generally have the most comprehensive water quality standards.

II. SETTING WATER QUALITY STANDARDS FOR DIFFERENT WATER USES

We have discussed standards that are set on such things as discharges and stream water quality but these are not the only standards used to regulate water quality. The most sophisticated and perhaps most important standards are those that regulate drinking water quality.

A. Drinking Water Quality Standards

The history of drinking water standards goes back thousands of years. Baker (1949) referred to a passage in a book written over 4000 years ago reflecting the need to boil water before consumption. The obvious purpose for establishing drinking water standards, of course, is to protect the public health. In particular, standards are set and treatment is instigated to prevent several diseases that are known to be waterborne.

The number of reported cases of waterborne disease has diminished significantly since the turn of the century. Figure 10.1 shows the total incidence of waterborne disease from 1920 to 1976. As treatment measures for drinking water were begun and stringent drinking water standards were developed, the rate of waterborne disease diminished. By 1950 only a few cases of waterborne disease were reported in the United States. However, after 1950 the number of waterborne diseases reported began to increase, and data from the late 1970s indicate that this trend is continuing. It is not clear whether the actual number of cases of waterborne disease is increasing or that the reporting of such disease is now more accurate. Although the total number of disease outbreaks is still in question, it is clear that the type of disease reported has changed in the last 30 years. Table 10.4 shows the etiology of waterborne disease outbreaks from 1946 to 1974. It is seen that gastroenteritis (of unknown etiology) represents by far the most frequently contracted waterborne disease. The deadly epidemics of typhoid and cholera that used to be dominant in the nineteenth century no longer

TABLE 10.3

| STANDARD | ILLINOIS | |
	POLLUTION CONTROL BOARD PROPOSED STANDARDS	SANITARY REVIEW BOARD STANDARDS
Biological standards		
BOD (mg l^{-1})	70	
Algae (ml^{-1})	500	
Fecal coliforms (100 ml)$^{-1}$	1000	5000
Fecal coliforms (primary contact) (100 ml)$^{-1}$	200	200
Threshold odor number		
Chemical standards (mg l^{-1})		
Ammonium nitrogen	1.0	2.5
Arsenic (dissolved)	0.01	0.05
Barium	1.0	1.0
Boron	0.01	0.01
Cadmium	0.01	0.01
Carbon chloroform extract	0.2	0.2
Chloride	250.0	150.0
Chromium (total)	0.05	
Trivalent	0.05	0.1
Hexavalent	0.05	0.05
Copper	0.02	1.0
Cyanide	0.01	0.025
Fluoride	1.4	1.0
Iron	0.3	0.3
Lead	0.05	0.05
Manganese	0.05	
Mercury (total)	0.0005	
Nitrates and nitrites (as N)	10.0	45 (as NO$_3$)
Oil (hexane solubles)	0.1	
Phenols	0.001	0.002
Phosphorus (as P)	0.1	4.0
Selenium	0.01	0.01
Silver	0.05	0.05
Sulfates	250	200.0
Zinc	5.0	5.0
Total solids	500	
average		500
maximum		750
Effluent treatment requirements	Specific schedules of BOD and suspended solid levels for various dilution ratios	Specific schedules of BOD and suspended solid levels for various dilution ratios

COMPARISON OF WATER QUALITY AND EFFLUENT TREATMENT
STANDARDS FOR TEN SELECTED STATES[a]

INDIANA (PUBLIC WATER SUPPLY)	IOWA (PUBLIC WATER SUPPLY)	MICHIGAN (DOMESTIC WATERS)
5000	2000	5000
200	200	1000
(3)		
0.05	0.05	0.05
1.0	1.0	1.0
0.01	0.01	0.01
		125 (avg.)
0.05	0.05	0.05
0.025	0.025	0.2
1.0	1.5	1.7
0.05	0.05	0.05
	0.02	0.002 (avg.)
0.01		
0.05		0.05
	None specified	
500		500
750		750
Secondary treatment	90% BOD and suspended solid reduction	Secondary treatment

(*Table continues*)

TABLE 10.3 (CONTINUED)

MINNESOTA (CLASS "C" WATERS)	MISSOURI (DRINKING WATER SUPPLY)	WISCONSIN (PUBLIC WATER SUPPLY)
4000	2000	5000
1000	200	1000
(3)		
0.01	There are minor variations in	The public water supply
1.0	water quality standards for	will be such that by
	different drainage basins in	appropriate treatment
0.01	the area	and adequate safeguards
0.2		it will meet the Public
250.0		Health Service Drinking
		Water Standards of 1962
0.05		
1.0		
0.01		
1.5	1.2	
0.3		
0.05		
0.05		
45 (as NO$_3$)		
0.001		
0.05		
250.0		
5.0		
	None specified	
		500
500		750
Reduction to 25 mg l^{-1} BOD, 30 mg l^{-1} suspended solids (not including algal cells), and 1000 coliforms/ 100 ml	Secondary treatment or 85% removal, whichever produces better effluent	90% removal of BOD and secondary treatment of suspended solids

OHIO (PUBLIC WATER SUPPLY)	PENNSYLVANIA (GROUP "A" WATERS)	NEW YORK (CLASS "A" WATERS)
5000	5000 (winter) 1000 (summer)	None specified
200	1000 (avg.)	
24 (at 60°C)	24 (at 60°C)	8
		2.0
0.05	0.05	
1.0		
0.01		0.3
	250.0	
0.05		
	0.1	0.2
0.025		0.1
1.0		
	0.3	
0.05		
		0.05
	0.3 (as PO_4)	
0.01		
0.05		
		0.3
		None specified
500	500	
750	750	
Secondary treatment	75 to 85% BOD reduction	Secondary treatment at all municipal plants

[a] Specific effluent requirements have not been adopted in the ten states with the exception of Illinois. Some states have specific stream quality standards for particular waters. The data shown indicate typical values for most waters.

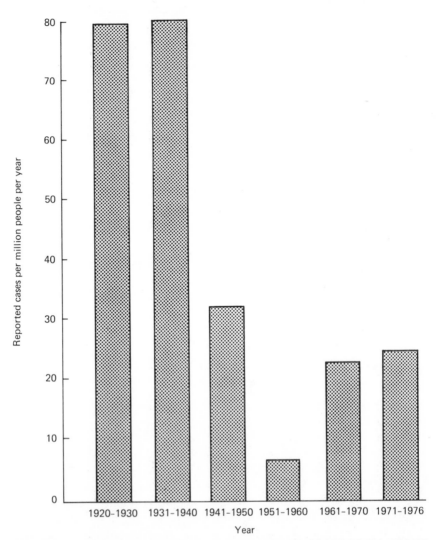

FIGURE 10.1   TOTAL INCIDENCE OF WATERBORNE DISEASE FROM 1920 TO 1976.   (From Craun, 1977.)

occur.   It therefore appears that strict drinking water standards, along with treatment requirements, have eliminated the more deadly waterborne diseases.   However, there is still a large incidence of gastroenteritis, and in many cases the causal agent is unknown.

It should also be noted that infectious hepatitis represents a large portion of reported waterborne diseases.   It is the only waterborne disease known to

TABLE 10.4   DISTRIBUTION OF ETIOLOGY BY WATERBORNE
DISEASE OUTBREAKS FROM 1946 TO 1974[a]

| DISEASE | OUTBREAKS FOR WATER SYSTEMS | | | |
| | COMMUNITY | | OTHER | |
	NUMBER	PERCENTAGE	NUMBER	PERCENTAGE
Gastroenteritis				
(unknown etiology)	71	52.2	153	47.6
Infectious hepatitis	22	16.2	44	13.7
Shigellosis	13	9.6	33	10.3
Chemical poisoning	8	5.9	13	4.0
Giardiasis	7	5.1	8	2.5
Typhoid	6	4.4	51	15.9
Salmonellosis	6	4.4	9	2.8
Amebiasis	1	0.7	4	1.2
Poliomyelitis	1	0.7	—	—
Enteropathogenic E. coli	—	—	4	1.2
Tularemia	—	—	2	0.6
Leptospirosis	1	0.7	—	—
Total outbreaks	136	—	321	—

[a] Sources: Craun and McCabe (1973) and Craun et al. (1976). © 1973, 1976 The
American Water Works Association.

be caused by a virus.   Although many studies have shown that viruses can
be isolated from sewage and surface waters, including poliovirus, echovirus,
and other animal viruses, very few documented cases of viral infections
from these agents can be found.

It appears that the use of drinking water standards with accompanying
treatment has reduced the incidence of waterborne disease in the United
States.   As noted earlier, there has been an increase in certain types of
waterborne disease, but the most deadly of the diseases have all but been
eliminated.   It is therefore useful to review the development of drinking
water standards in the United States and see what standards are currently
followed for the treatment of drinking water.

Drinking water standards in the United States originated in the Congress
of 1893, which passed the Interstate Quarantine Act.   This act gave the
Surgeon General power to take any measures necessary to prevent the
transmission of communicable diseases.   As a result, in 1912 a regulation
was adopted that prevented the use of common drinking cups on interstate
water carriers.   The first actual standard was issued in 1914 when a bacteriological standard was adopted for water transported on interstate carriers.

Although these standards were only applicable to interstate water carriers, many state and local agencies adopted similar standards.

In 1925 the standards were revised and new sections were added that set limits on various chemical and physical constituents of water. The 1925 standards included limits on lead, copper, zinc, and other soluble minerals.

In 1942 the standards were again revised and maximum permissible concentrations for various chemicals such as barium, chromium, arsenic, and selenium were established. In addition, the 1942 revisions included monitoring requirements throughout the entire interstate water distribution system. The standards mandated that the entire system be free from any defects that might cause contamination of the water.

In 1946 the standards were again revised, principally to make the requirements applicable to all water supplies in the United States. Very little with respect to content of the standards was changed at this time.

In 1962 the standards were revised once again, and the revision committee recommended that many of the limits be based on the volume of water consumed. This meant that climate would have some impact on the quality standards of water allowed for human consumption. The 1962 standards also included standards for radioactivity.

Between 1962 and 1973 several advisory committees recommended changes, principally additions, to the Federal Water Quality Standards. These revisions included limits for toxic chemicals such as pesticides and other toxic pollutants. Several of these changes have become water quality standards, whereas others are just recommended limits. In 1973 the standards were once again reviewed and updated. These new water quality standards were adopted in the late 1970s, and their influence can be seen in many state and local drinking water standards. The basic allowable limits of most chemical constitutents of water systems remain the same as set forth in the 1962 standards. The trend has been to allow state and local communities to regulate their own drinking water quality. In most cases these agencies adopt the federal drinking water standards, but in some cases they may set more stringent standards on certain chemical pollutants. Table 10.5 shows typical finished water quality standards for a public water supply. The table refers to standards issued by the state of Illinois, and they are essentially identical to federal drinking water standards.

As mentioned previously, drinking water standards are based on medical studies and epidemiological evidence that establish safe pollutant levels from a public health standpoint. Most communities (states, counties, regions, etc.) adopt federal standards because of their sound medical basis. As new medical evidence becomes available, the standards are adjusted accordingly. Epidemiological evidence to date shows that these standards are effective, because very few incidences of disease have occurred in the

TABLE 10.5 TYPICAL FINISHED WATER QUALITY STANDARDS BY MAXIMUM
ALLOWABLE TWELVE-MONTH-AVERAGE CONCENTRATIONS

SUBSTANCE	REPORTED AS	MAXIMUM CONCENTRATION (mg/l)	COMPLIANCE DATE
Arsenic	As	0.1	E[b]
Barium	Ba	1.0	Jan. 1, 1978
Cadmium	Cd	0.010	E
Chromium	Cr	0.05	E
Color	Color units	15.0	E
Copper	Cu	1.0	E
Cyanide	CN	0.2	E
Fluoride	F	2.0	Jan. 1, 1978
Foaming agents	MBAS[a]	0.5	E
Iron	Fe	0.3	Jan. 1, 1978
Lead	Pb	0.05	E
Manganese	Mn	0.05	Jan. 1, 1978
Mercury	Hg	0.002	E
Nitrate nitrogen	N	10.0	E
Nitrite nitrogen	N	1.0	E
Odor	Threshold odor number	3.0	E
Organics			
Adsorbable carbon			
Carbon–chloroform extract	CCE_m	0.7	E
Pesticides			
Chlorinated hydrocarbon			
insecticides			
Aldrin		0.001	E
Chlordane		0.003	E
DDT		0.05	E
Dieldrin		0.001	E
Endrin		0.0005	E
Heptachlor		0.0001	E
Heptachlor epoxide		0.0001	E
Lindane		0.005	E
Methoxychlor		0.1	E
Toxaphene		0.005	E
Organophosphate			
insecticides			
Parathion		0.1	E
Chlorophenoxy herbicides			
2,4-Dichlorophenoxyacetic			
acid (2,4-D)	0.02	E	
2,4,5 Trichlorophenoxypro-			
prionic acid	0.01	E	
(2,4,5-TP or Silvex)			
Selenium	Se	0.01	E
Silver	Ag	0.05	E
Turbidity	Turbity units	1.0	E
Zinc	Zn	5.0	E

[a] MBAS, methylene blue active substances.
[b] E, effective date of rules and regulations.

United States since the standards have been enforced. As new compounds, especially toxic industrial compounds, are isolated from public water supplies, new standards have to be created. Therefore, the establishment of standards is a dynamic phenomenon, changing through the years.

Standards established for drinking water quality are perhaps the easiest to set. This is because estimates of daily water intake and medical studies relating the effects of certain contaminants to dosage are easy to correlate. Water quality standards for uses other than drinking are more difficult to establish. In such cases a factor beyond the direct consumption of water must be established, and this problem will be exemplified in the following sections.

B. Recreational Water Quality Standards

Society now demands that water systems, especially surface waters, be free from hazardous compounds and disease-causing organisms. In addition, surface water quality must be maintained at an acceptable level for body contact as well as being maintained at a level sufficient for effective drinking water treatment. Standards must therefore be established to protect surface waters from degradation. The concern for maintaining high-quality receiving waters means that unit processes must be included for the treatment of waste streams. The treatment of wastewater is very expensive, and therefore reasonable standards must be set that take into account the degree of treatment required. These standards must protect the desired biological system of the receiving waters and make them safe for human contact.

Perhaps the most important recreational use of water, with respect to setting standards on quality, is swimming. Swimming involves direct body contact and therefore the greatest chance of contamination from pollutants among recreational uses. Establishing the incidence of disease transmission from body contact, however, is difficult. In addition, relating water-caused illness to the overall quality of the water is also difficult. This is because of the background level of illness in any community (which may change cyclically with the seasons) that can disguise the effect of contaminated water. Many studies have attempted to evaluate the impact of swimming in polluted waters on human health, but conflicting results have been obtained.

Cabelli and McCabe (1974) attempted to define the hypothetical water quality criteria for recreational waters. Figure 10.2 shows their approach to the problem. It can be seen that, depending on the quality of the water, there will be a certain incidence of both mild and severe illness. The mild

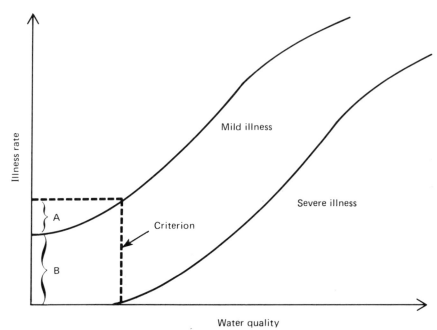

FIGURE 10.2   A HYPOTHETICAL WATER QUALITY CRITERION FOR RECRE-
ATIONAL WATERS.   Illness rates can be differentiated by type (respiratory, gas-
trointestinal, etc.) or by severity (stomachache, restricted activity, or required medical care).   Water quality can be measured by the density of indicator organisms or
specific pathogens.   A, acceptable risk; B, background illness.   (From Cabelli and
McCabe, 1974.)

illness rate does *not* approach zero at a very high water quality, representing
the background illness rate.   Therefore, there should be some acceptable
level of illness that possibly precludes any severe illness due to water
contact.   This designed acceptance of a slightly increased illness rate would
then establish the water quality standard for the water system.   This approach is highly theoretical, and we will see later that it is difficult to
establish both the illness rate due to water contact and the acceptable illness
rate, both necessary for developing a standard.

An evaluation of the types of diseases that are contracted by water contact
is important for ascertaining the effect of polluted waters.   Hollis *et al.*
(1976) reported that the most common infections associated with swimming are those of the skin, ears, eyes, and upper respiratory tract.   In a more
comprehensive study, Stokes (1927) collected data on the number of cases
of various infections that occurred among swimmers.   Table 10.6 shows the
relative occurrence of various infections in a swimming population.   It can

TABLE 10.6   OCCURRENCE OF INFECTION
AMONG SWIMMERS[a]

TYPE OF INFECTION	CASES	PERCENTAGE OF CASES
Eye	150	30.1
Ear	148	29.7
Upper respiratory	106	21.3
Throat	13	2.6
Skin	81	16.3

[a] From Stokes (1927).

be seen that eye and ear infections are the most common infection contracted by swimmers.

More serious infections are contracted by swimmers at a much lower rate. For example, leptospirosis can be contracted by people swimming in waters contaminated with the urinary wastes of cattle. The most serious illness associated with swimming, however, is primary amoebic meningoencephalitis, which is caused by members of the genus *Naglaria.* This disease is almost always fatal, and fortunately very few cases are reported in the United States. The organisms involved are generally associated with freshwater systems in the southern United States and are predominant during the warm summer months. This particular illness has been of concern in areas where thermal effluents are discharged, raising the ambient temperature of the surface waters. Because of the possible increase of this particular disease, controls are being initiated to regulate the temperatures of surface waters more closely.

Other illnesses contracted by people swimming in contaminated water have been reported, but at very low rates. As an example, gastroenteritis has been reported at several recreational facilities, but the incidence is very low. Also, diseases such as typhoid fever and poliomyelitis are reported to occur very infrequently.

As mentioned previously, designating polluted water as dangerous to swimmers is very difficult, and studies to evaluate the incidence of disease in bathers swimming in contaminated water have yielded mixed results. Table 10.7 shows illness rates by type at selected New York City beaches. It is known that some of the beaches are more polluted than the others tested in the study. It should be noted that there is a difference in symptom type among the beaches, indicating that water quality does in fact affect the illness rate of swimmers.

The possibility of infection of the skin, ears, and the upper respiratory tract apparently increases as the quality of the water deteriorates. The

TABLE 10.7   ILLNESS RATES BY TYPE AT CONEY
ISLAND AND ROCKAWAYS BEACHES[a]

ILLNESS CATEGORY	ILLNESS RATE (%)	
	CONEY ISLAND	ROCKAWAYS
Respiratory	2.7	6.3
Gastrointestinal	4.8	3.5
Other	3.3	0.5
Severe symptom(s) (incapacitating or relatively serious)	1.7	0.4

[a] From Cabelli and McCabe (1974).   Values given are the rates for swimmers less the rates for nonswimmers.

reported incidence of severe illness caused by swimming, however, is extremely low.   It therefore appears that the quality of water must be severely deteriorated by recent waste discharges for such a significant increase in water-contact illness to occur.   However, because there is some level of illness associated with contaminated water, standards must be established to minimize the incidence of disease.   We shall now address the types of standards established for the protection of water quality for recreational use.

Because most water-contact illnesses contracted by humans are of human origin, most recreational standards regulate human contamination.   The most common indicator of fecal contamination of surface waters is the coliform group.   Coliforms are discussed in Chapter 6, along with methods for analyzing their occurrence.   It is clear that if a water system has recent contamination of fecal origin, then the possibility of disease transmission exists.   This possibility increases when the fecal matter of a large portion of the population contains organisms capable of causing specific diseases.   Therefore, the use of coliform bacteria to identify fecal pollution in recreational waters seems appropriate.   The question of establishing standards, that is, determining the acceptable level of microorganisms, becomes a more difficult task.

The assumption that fecal contamination is the only cause of water-contact diseases may be inaccurate.   It has been shown that pathogenetic organisms can exist in the fecal matter of livestock as well as of domestic pets.   Many of these microorganisms are pathogenic to man and can contaminate fish that are eventually consumed by humans.   This problem exists with microorganisms such as *Salmonella* and *Leptospira*.   Therefore, standards utilizing coliform bacteria must take into account the fact that contamination from animals other than humans can also occur.

The establishment of an acceptable indicator level is important when setting recreational standards. The use of only coliform bacteria as an indication of contamination may not be effective in regulating water quality. Geldreich (1970) showed that the correlation between the presence of fecal coliforms with the occurrence of *Salmonella* is very good. Table 10.8 shows data from studies that compared the occurrence of fecal coliforms to the occurrence of *Salmonella* in both freshwater and estuarian systems. The data indicate a sharp increase in the frequency of detection of *Salmonella* when fecal coliform levels are greater than 200 organisms ml^{-1}. When the fecal coliform density exceeds 2000 organisms ml^{-1}, isolations of *Salmonella* can be made with nearly 100% assurance. These studies indicate that a standard of aproximately 200 fecal coliform organisms ml^{-1} would be a reasonable standard for protecting recreational waters.

Theoretically, the establishment of microbiological standards to protect those people in contact with water supplies should be set such that no health risk exists. There have been several microbiological standards established for recreational waters around the country, and these generally use values or limits for total coliform bacteria ranging from 0.5 to 30 organisms ml^{-1}. Although the total coliform count may not be an accurate measure of fecal contamination, there are correlations that show marked decreases in the incidence of illness when total coliform levels are less than about 10 organisms ml^{-1}. Because of the discrepancy between the occurrence of fecal coliforms and total coliforms, recreational use of water systems may be unduly restricted if standards are based on coliform counts. For instance, if a situation arises in which a large number of coliform microorganisms are

TABLE 10.8 THE CORRELATION OF FECAL COLIFORM WITH *SALMONELLA* OCCURRENCE[a]

SOURCE	FECAL COLIFORM DENSITY (per 100 ml)	*SALMONELLA* DETECTION		
		TOTAL EXAMINATIONS	POSITIVE OCCURRENCE	
			NUMBER	PERCENTAGE
Freshwater	1–200	29	8	27.6
	201–2000	27	19	85.2
	over 2000	54	53	98.1
Estuarine water	1–70	184	12	6.5
	71–200	74	21	28.4
	201–2000	91	40	44.0
	over 2000	75	45	60.0

[a] From Geldreich (1970). Reprinted from *J. Amer. Water Works Assoc.*, Vol. 62, No. 2 (February 1970), by permission. © 1970 The American Water Works Association.

isolated, recreational facilities may be closed to prevent any health risk to the population. This situation could arise because of stormwater runoff, which is known to contain large concentrations of total coliforms, although this would not necessarily indicate fecal contamination of the water system. Therefore, the use of *fecal* coliforms, rather than *total* coliforms, as an indication of recreational water quality is probably more directly related to the actual degree of fecal contamination.

In conclusion, the use of microbiological indicators for recreational water quality, although conceptually useful, varies in effectiveness in practice. In most cases the current practice is to set a limit on total coliforms, for example, 10 microorganisms ml^{-1}, to protect swimmers from contaminated water. Establishing the standard implies that when the population of indicator microorganisms exceeds the acceptable level, the facilities must be closed. In many municipal areas in the country closing recreational facilities means social hardship and loss of income for the community. Therefore, the establishment of accurate and effective standards is very important. However, recreational standards, by their nature, cannot be as precise as microbiological standards for drinking water quality. In the latter case the occurrence of any coliforms at all means that the water may be unsafe for human consumption and requires further treatment. However, using microbiological standards for recreation means that some level or acceptable concentration of microorganisms can be allowed. The establishment of the exact level necessary to protect human health without unduly restricting use of the resource is a difficult task.

III. WATER QUALITY DEGRADATION FROM LAND USE PRACTICES

The most common sources of pollution and perhaps the most easily recognizable are those designated as point sources. These are composed of discharges from domestic waste treatment facilities and industrial complexes. The degree of treatment of these effluents has been the subject of years of study and many approaches are now available for rendering such waste streams innocuous. The ultimate goal of zero discharge exemplifies the efforts that have been made to stop inputs of pollution to water systems in the United States. However, it has been estimated that perhaps 80% of the urban areas in the United States will *not* realize increased water quality from extensive treatment of point sources, that is, the water quality downstream from these urban areas is not controlled by pollution inputs from point sources. Recent evidence indicates that pollution inputs from nonpoint sources in the form of stormwater runoff or drainage due to routine land use practices is responsible for a large part of the degradation of surface water quality.   Indeed, rainwater itself is now heavily contaminated in many areas by airborne pollutants (see Chapter 8).   Rainfall has been shown to contain many solid and dissolved components that are deposited in surface water systems during rainstorms.

Urban runoff contains organic wastes, bacterial contaminants, metals, suspended solids, and other pollutants that are entrained in the water as it runs over ground surfaces.   The effect of stormwater runoff on surface water quality is therefore an important consideration for communities that are investing large sums of money in sewage treatment facilities.

One of the first studies to evaluate the impact of stormwater runoff on surface water quality was performed in North Carolina (Bryan, 1970).   This study showed that the biochemical oxygen demand (BOD) of stormwater runoff was essentially equivalent to that of domestic wastewater effluent following secondary treatment.   The chemical oxygen demand (COD) in

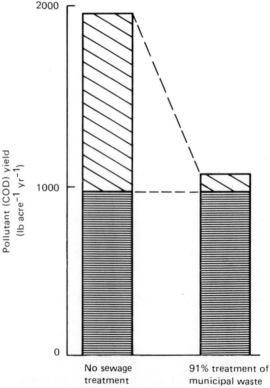

FIGURE 10.3   COMPARISON OF THE ANNUAL POLLUTANT (COD) YIELDS FOR MUNICIPAL SEWAGE (▨) AND URBAN RUNOFF (▤).   (From Bryan, 1970.)

the stormwater was actually greater than that attributed to any discharge of raw sanitary wastewater from an equivalent urban area. In like manner, the total contribution of suspended solids from stormwater was in excess of that expected from the average raw domestic wastewater. Figure 10.3 shows a comparison of municipal sewage and urban runoff with respect to pollution potential. Even with greater than 90% treatment of municipal sewage, less than half of the total pollutant yield (measured as COD) would be removed. For any substantial decrease in COD to be observed in this situation, there would have to be some sort of treatment of the urban stormwater runoff. This could conceivably be achieved by the containment and sedimentation of the urban runoff or by actual chemical treatment.

Although the North Carolina study indicated large inputs of pollutants, especially in the form of BOD, COD, and suspended solids, from urban runoff, the input of nutrients was relatively small. For example, the

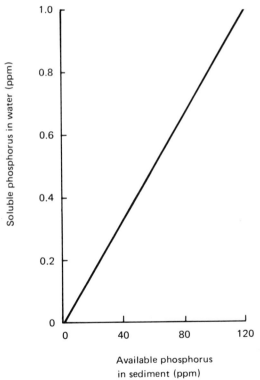

Available phosphorus
in sediment (ppm)

FIGURE 10.4 RELATIONSHIP BETWEEN SOLUBLE ORTHOPHOSPHATE CONCENTRATION IN RUNOFF WATER AND AVAILABLE (I.E., EXTRACTABLE) PHOSPHORUS IN RUNOFF SEDIMENT. (From Römkens and Nelson, 1974.)

contribution of phosphorus from urban stormwater was found to be nominal when compared to that from domestic wastewater. This situation evolves because of the impervious nature of the surfaces in most urban areas. The concentration of nutrients in runoff, of course, depends on the concentration of nutrients in rainfall, fertilizer usage in the area, and the degree of contact between stormwater runoff and soils containing nutrients.

An evaluation of phosphorus addition due to runoff from fertilized soils was attempted in a study by Römkens and Nelson (1974). Figure 10.4 shows that there is a definite correlation between the amount of phosphorus solubilized and the amount of phosphorus in the sediment. As the phosphorus content of the sediments increases, so does the amount extracted by stormwater. The actual amount of phosphorus that can be leached from soil, however, depends on the chemical composition of the soil. Soils containing large amounts of aluminum or iron bind phosphorus more tightly and hence lose less to the runoff. This solubilization of nutrients, especially phosphorus, tends to enrich surface waters and subsequently aggravate the eutrophication process.

Perhaps the most inclusive study that considers stormwater pollution from urban runoff was performed by the Environmental Protection Agency (1970) in Tulsa, Oklahoma. In this study stormwater runoff samples were collected from 15 different test areas in the Tulsa metropolitan area and analyzed for BOD, COD, TOC, organic nitrogen, phosphorus, chloride, pH, suspended solids, coliforms, and streptococci. A brief description of each of the test areas is shown in the accompanying list.

Test area	Description
1	A new industrial district
2	Shopping center
3	Middle-class residential neighborhood
4	Industrial, commercial, and residential use
5	Fully developed residential area
6	Mixture of industrial and residential activity
7	Entirely residential area
8	Residential area in which streets are not guttered
9	Lower-class residential area
10	Mixture of commercial and residential land uses
11	Large drainage basin, residential and commercial
12	Airport drainage
13	Residential, expensive homes on large tracts
14	Golf course drainage
15	Residential area with small homes

A summary of the findings of this study is given in Table 10.9, which shows a yearly average load factor for the pollutants investigated in each of the areas. The first and most obvious fact is that substantial load from all

TABLE 10.9 CALCULATED AVERAGE YEARLY LOADS OF
POLLUTANTS IN STORMWATER RUNOFF FROM 15 TEST AREAS IN
TULSA, OKLAHOMA[a]

| | | | | POLLUTION LOAD (lb acre^{-1} yr^{-1}) | | |
TEST AREA	ACRES	BOD	COD	ORGANIC KJELDAHL NITROGEN	SOLUBLE ORTHOPHOSPHATE	TOTAL SOLIDS
1	686	30	250	2.5	8.0	5100
2	272	27	150	3.3	2.9	920
3	550	14	110	2.6	3.3	1200
4	938	44	320	3.0	3.3	1900
5	507	33	250	1.3	1.6	490
6	368	21	160	1.1	1.5	600
7	197	15	90	1.5	1.3	790
8	211	33	250	1.5	2.5	840
9	64	20	230	1.3	2.0	830
10	206	48	470	3.6	3.1	1900
11	815	35	290	1.7	2.1	1400
12	223	25	140	1.2	1.7	630
13	212	25	150	2.4	2.0	780
14	263	12	60	1.1	1.1	660
15	74	25	90	0.8	1.7	570

[a] From Environmental Protection Agency (1970).

forms of land use are witnessed for all of the common pollutants. There are
substantial loads per acre of COD, total solids, and soluble orthophosphate. There is some variation among land use practices with respect to the
addition of specific pollutants, but they all contribute substantial amounts
of each pollutant. The total loads given in this table are based on the
average annual rainfall in the Tulsa metropolitan area. Pollutant loads will
obviously be different in different parts of the country depending on the
amount of rainfall; however, the data do indicate that urban runoff has a
substantial effect on surface water quality.

IV. WATER QUALITY CONTROL MEASURES

A. Developing a General Model

It is assumed that the water quality of any system, whether fresh, brackish,
or marine, depends directly on the amount of waste delivered to it. In

general, relatively small waste loads mean that very little environmental damage is inflicted. As waste loads increase, the detrimental effects also increase. Figure 10.5 is an ideal representation of changes in water quality as a function of waste load, showing that there is a band of possible responses to the water quality of a system depending on the type and quntity of waste inputs. We have already defined waste inputs as either point sources or non-point sources and have evaluated both types. The types of waste inputs will now be defined in accordance with the National Research Council.

(1) *Oxygen-demanding materials* include all organic wastes from domestic and nondomestic sources. Under aerobic conditions all of these organic materials require oxygen for decomposition and are therefore considered to be oxygen-demanding wastes. The rate at which oxygen is consumed depends on the specific waste and is the basis for determining the amount of treatment required or the degree of oxygen depletion in receiving waters.

(2) *Nutrients* include the macronutrients required for algal growth, such as nitrogen, phosphorus, silica, and other minor elements. We have seen that domestic waste usually includes high quantities of all of these nutrients, especially nitrogen and phosphorus.

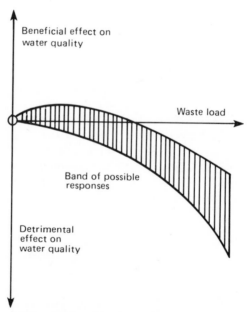

FIGURE 10.5 AN IDEALIZED REPRESENTATION OF WATER QUALITY AS A FUNCTION OF WASTE LOAD.

(3) *Pathogens and carcinogens* are disease-causing agents or organisms, including bacteria, viruses, protozoa, and chemical mutagens. Domestic sources are responsible for many of the pathogens, and industrial and agricultural sources usually generate most of the carcinogens.

(4) *Radionuclides* include all artificially produced radioisotopes. Radionuclide wastes can be generated from mining or refining activities in the industrial sector.

(5) *Inert material* includes sediment, inorganic materials, organic materials with very slow rates of decay, and salts or brine wastes. These wastes are generated in both the domestic and industrial sectors.

(6) *Oils and grease,* including scums and particulates from petroleum residues, are usually discharged to the environment by accidents during the transportation of the product.

(7) *Thermal pollution* generally comes from power generation.

(8) *Poisons* include hydrocarbons such as DDT and PCB as well as other organic chemicals. Most insecticides fall within this group. Several metallic poisons can also be classified under this heading.

The relative contribution of each of these pollutants varies with time and space, but in most cases they are all discharged to some extent. To determine the management options available for dealing with excessive waste loads, the waste discharges themselves must be defined. Wallis (1973) proposed a general mathematical formula for evaluating waste discharges to a given watershed. He divided the watershed into k sections with the generation of wastes defined as

$$G_{ijk} = \sum_{i=1}^{6} P_k(t)C_{ijk}(t), \tag{1}$$

where G is the quantity of waste generated, P the population, C the average per capita generation of wastes, i the summation over all waste sources, j the waste type, and k the section number. If sewage treatment plants are located at L points in the watershed and the transport of waste from section k to a treatment process located at plant ℓ can be represented by $Z_{k\ell}$, then the removal efficiency of each treatment plant can be expressed as

$$T_{j\ell}(t) = \sum_{k=1}^{k} \eta_{ij}(t)G_{jk}(t)Z_{k\ell}, \tag{2}$$

where $\eta_{j\ell}$ is the waste treatment efficiency for waste type j at plant ℓ. After treatment the wastes are transported to the receiving water $Z_{\ell m}$ at section m. The discharge of the waste is therefore

$$Q_{jm}(t) = \sum_{i=1}^{G} \sum_{k=1}^{K} \sum_{\ell=1}^{L} [1 - \eta_{j\ell}(t)]P_k(t)C_{ijk}(t)Z_{k\ell}Z_{\ell m}. \tag{3}$$

After the waste has been discharged from the treatment plants, it is distributed in space and time according to

$$Z_{jk\ell mn} = Z_{k\ell}Z_{\ell m}Z_{jmn}, \tag{4}$$

where Z_{jmn} is a waste distribution function between sections in the receiving water. The assimilative capacity of the receiving water is considered as an alteration of the waste distribution function. This is achieved by multiplying by $1 - A_{jmn}(t)$, where $A_{jmn}(t)$ represents a reduction in waste type j owing to increased assimilation capacity in section m.

Combining Eqs. (3) and (4), a general expression for the waste load in a receiving water R is developed:

$$R_{jn}(t) = \sum_{i=1}^{6} \sum_{k=1}^{K} \sum_{\ell=1}^{L} \sum_{m=1}^{M} [1 - \eta_{ij\ell}(t)]P_{jik}(t)$$

$$\times C_{ijk}(t)Z_{jk\ell mn}(t)[1 - A_{jmn}(t)]. \tag{5}$$

The options available can be obtained by differentiating Eq. (5):

$$\frac{dR(t)}{dt} = -\frac{d\eta}{dt}\bigg|_{PCZA} PCZA(1 - A) + \frac{dP}{dt}\bigg|_{\eta CZA}(1 - \eta)CZ(1 - A)$$

$$+\frac{dC}{dt}\bigg|_{\eta PZA}(1 - \eta)PZ(1 - A) + \frac{dZ}{dt}\bigg|_{\eta PCA}(1 - \eta)PC(1 - A) \tag{6}$$

$$-\frac{dA}{dt}\bigg|_{\eta PCZ}(1 - \eta)PCZ.$$

It is clear from Eq. (6) that the options available for reducing wastes inputs to water systems are

(1)   increasing the removal of waste during treatment,
(2)   reducing the population that generates the waste,
(3)   reducing the per capita generation of waste,
(4)   changing the distribution of waste,
(5)   increasing the assimilative capacity of the aquatic system.

It can be seen from the first term on the right-hand side of Eq. (6) that for the overall waste generation to be minimum $d\eta/dt$ must be greater than zero. This means that there should be some procedure developed to remove waste during treatment or to increase the efficiency of the waste treatment process. The second term on the right-hand side represents a change in the waste-generating population over time. For the total waste generation term to be less than zero dP/dt must also be less than zero. Therefore, a decrease in waste load can be achieved if the growth of population in certain areas is restricted, but the means for achieving this are beyond the scope of this text.

The third term on the right-hand side of Eq. (6) represents a change in per capita waste generation, and again dC/dt must be less than zero for the overall waste load term to be small, that is, a decline in the waste load over time could be achieved by reducing the per capita generation of waste. This option implies that alternative industrial procedures should be initiated, including a way of minimizing the use of raw minerals.

The last two terms on the right-hand side of the equation deal with the distribution of waste. Equation (6) indicates that the total waste load can be changed by either redistributing the deposition of the waste material or increasing the assimilative capacity of the receiving system. These two components of the model can be addressed as they relate to water quality management options. In particular, two approaches seem feasible for implementing these considerations: the diversion of waste flows from sensitive areas and the engineering of natural systems to either remove pollutants *in situ* or increase the assimilative capacity of the receiving system. We shall address these options in the following sections, because recent work has indicated that they might be feasible alternatives for minimizing water quality deterioration due to waste inputs.

B. Wastewater Diversion

The diversion of pollution sources is always the initial step in any water restoration scheme. There are many projects on record that have diverted waste from lakes, streams, or estuaries in an effort to upgrade water quality. Most of these projects diverted domestic wastes to minimize nutrient inputs and therefore slow the eutrophication process. In Chapter 5 we discussed the effect of nitrogen and phosphorus inputs on natural water systems, and it was noted that domestic waste accounts for over 50% of the nutrients added to aquatic systems.

The diversion of wastewater from lakes to prevent massive algal growth has been attempted with varying degrees of success. Wastewater was diverted from Lake Lyngby-Sø in Denmark in 1959, and the lake recovered very rapidly. On the other hand, a lake restoration project in Wisconsin showed that Lake Monona was very slow to recover from wastewater diversion, even after 34 years (Lee, 1966). Red Lake in Switzerland, which underwent wastewater diversion in 1933, has never shown satisfactory recovery. The eutrophication level of the lake has remained constant even though the wastewater input has been absent for many years.

Perhaps the best-documented study of this variability in the responses of lakes to restoration by nutrient diversion was undertaken for two lakes in Washington, one of which was Lake Washington in Seattle. This lake has been studied extensively by Edmondson (1970), and its eutrophication level

TABLE 10.10 PREDIVERSION CONDITIONS IN LAKE SAMMAMISH AND LAKE WASHINGTON[a]

PARAMETER	LAKE SAMMAMISH 1964–1965	LAKE WASHINGTON 1962–1963
Total P (μg l)		
Surface mean, October to April	24	61
Surface maximum	190	79
Nitrate N (μg l)		
Surface mean, October to April	142	323
Surface maximum	330	480
Chlorophyll a (μg l)		
Surface mean, annual	4.5[b]	20[c]
Surface maximum	27	45
Phytoplankton		
Surface volume maximum (μm^3 ml^{-1})	4.0×10^6	$14 \times 10^{6 d}$
Percent Myxophyceae	72	98[d]
Filamentous Myxophyceae		
Surface maximum (μm^3 ml^{-1})	1.3×10^6	6.0×10^6
Seston surface maximum, dry weight, uncombusted (mg l^{-1})	9.5	9.8
Secchi disk (m)		
Summer mean	2.9	1.2
Maximum	5.1	5.7
Minimum	2.0	0.9
Oxygen deficit, 2-yr mean (mg day^{-1} cm^{-2})	0.053	0.055
Oxygen production, maximum (g day^{-1} m^{-2})	3.7	7.0[d]

[a] From Emery *et al.* (1973).
[b] Averaged over 1965.
[c] Averaged over 1963.
[d] Highest observed for any year (1964).

was monitored for many years after the diversion of secondary effluent from 10 wastewater treatment plants in metropolitan Seattle. The study showed quite clearly that within 3 to 4 years after the wastewater effluent was diverted Lake Washington recovered in a remarkable manner. Lake Sammamish, which is located approximately 4 miles from Lake Washington, also underwent nutrient diversion but did not recover in the same manner. There was an extensive study by Emery, Moon, and Welch (1973) to determine why two lakes in very close proximity differed in their apparent recovery rates after wastewater diversion. Table 10.10 shows the prediversion conditions in the two lakes. Note that Lake Washington was in much worse shape before diversion than was Lake Sammamish. All of the common parameters that indicate trophic level, that is, total phosphorus, total nitrogen, and chlorophyll level, show that the eutrophication of Lake Sammamish was not nearly as advanced as that of Lake Washington. It should also be noted at this point that Lake Sammamish is much smaller and shallower than Lake Washington. The effect of the volume of hypolimnic water is an important consideration in predicting the final water quality of a lake after nutrient diversion.

Table 10.11 shows the effects of diversion on the two lakes. It can be noted that the phosphorus load to Lake Washington dropped from 1.06 to 0.48 gm yr^{-1} m^{-2}, which is a reduction of approximately 55%. On the other hand, the diversion of wastewater from Lake Sammamish reduced the phosphorus load from only 0.97 to 0.6 gm yr^{-1} m^{-2}. However, even though the rate of phosphorus addition to Lake Sammamish was somewhat reduced, the surface concentration of phosphorus was not diminished after 4 years. There is therefore no reason to expect any reduction in algal productivity or biomass, because the aqueous concentration of nutrients was not diminished. This continued degradation of water quality, even after waste diversion, is demonstrated in Fig. 10.6 by a comparison of chlorophyll levels in 1965 and 1970. It can be seen that even after waste diversion, no change in the algal biomass in the lake was observed.

There are several reasons why this phenomenon developed. The fact that Lake Sammamish is much shallower means that there is a greater retention of phosphorus in the bottom sediments and probably a greater regeneration rate of phosphorus than is the case for Lake Washington. The estimated benthic area exposed to hypolimnic waters for Lake Sammamish was 5.003 km^2, with the hypolimnic volume being 0.174 km^3, whereas that for Lake Washington was 20.15 km^2, with the hypolimnic volume being 1.15 m^2. The ratio of hypolimnic benthic area to hypolimnic volume for Lake Sammamish is then 28.7, whereas that for Lake Washington is only 17.28. This means that Lake Sammamish has approximately 65% more bottom sediment exposed to the hypolimnic water than does Lake Washing-

TABLE 10.11 COMPARISON OF DIVERSION EFFECTS ON LAKE SAMMAMISH AND LAKE WASHINGTON[a]

PARAMETER	LAKE SAMMAMISH	LAKE WASHINGTON
Area (km²)	19.8	87.615
Volume (km³)	0.350	2.884
Maximum depth (m)	31	64
Mean depth (m)	17.7	32.9
Flushing time (yr)	4.8	3.2
Prediversion annual total P income (kg)	19,100	92,600
Prediversion annual total P income per surface area (g yr⁻¹ m⁻²)	0.97	1.06
Percentage total P income diverted	39	≥55
Postdiversion annual total P income (kg)	11,800	41,700
Postdiversion annual total P income per surface area (g yr⁻¹ m⁻²)	0.60	0.48
Vollenweider's danger limit of P loading for respective mean depth (g yr⁻¹ m⁻²)	0.26	0.42
Prediversion annual inorganic N income (kg)	49,100	246,100
Prediversion annual total N income per surface area (g yr⁻¹ m⁻²)[b]	4.96	5.63
Percent inorganic N diverted	22	12
Postdiversion annual inorganic N income (kg)	38,298	216,568
Postdiversion annual total N income per surface area (g yr⁻¹ m⁻²)[c]	3.87	4.33
Vollenweider's danger limit of N loading for respective mean depth (g yr⁻¹ m⁻²)	4.00	6.00

[a] From Emery et al. (1973).
[b] Total N values are estimated by doubling inorganic N values.
[c] Estimated on the basis of population equivalent nutrients diverted and prediversion annual income to the lake.

ton, which indicates that there is a relatively greater transfer surface for the movement of phosphorus between sediments and overlying waters.

There is also the possibility that not enough phosphorus was precluded from the lake system by the diversion of domestic wastewater. The phos-

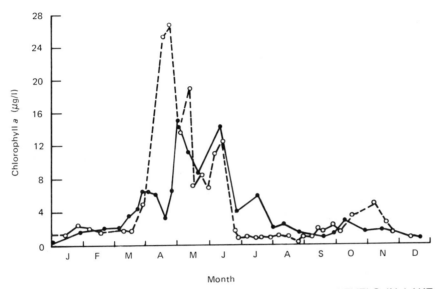

Month

FIGURE 10.6   ANNUAL VARIATION IN CHLOROPHYLL *a* LEVELS IN LAKE SAMMAMISH, SEATTLE, WASHINGTON, FOR THE YEARS 1965 (○) 1970 (•). (From Emery *et al.* 1973.)

phorus load rate to Lake Sammamish after diversion was still considerably higher than values necessary to limit excessive algal growth. The comparison of these two lakes shows that wastewater diversion (i.e., nutrient diversion) may not always be an effective means of restoring lake water quality. Obviously, many factors have to be considered, including the actual limiting nutrient, the location of the lake, the actual percentage of phosphorus or nitrogen that can be removed by diversion, and the trophic level of the lake before diversion.

C. Impoundment Aeration

The classical pattern of thermal stratification in temperate lakes is well understood. Temperature gradients, which develop in lakes as a function of changes in climate, effectively separate lakes into different bodies of water based on differences in density. This partitioning is caused by a thermocline that effectively separates the bodies of water. The thermocline is most pronounced during the summer months when the denser cold water remains at the bottom of the lake (the hypolimnion) and the less dense warm water remains above the thermocline (the epilimnion). The epilimnic waters are mixed during the warm summer months by wind currents, and any nutrient inputs to the lake, for example, surface runoff, fertilize the epilimnic waters. Large algal growths subsequently occur in the epilimnion because of the availability of sunlight, good wind circulation, and the surface water input of nutrients. The hypolimnic waters, on the other hand, remain stagnant, have no circulation, and receive organic and inorganic particulate material that settles from the epilimnion.

The stratification that evolves in temperate lakes affects many parameters of the lake system, but most importantly it affects the dissolved oxygen system. (The importance of dissolved oxygen as a basic biological and chemical parameter within a lake was discussed in Chapter 2, Section V.) Dissolved oxygen is mobilized in lakes by photosynthesis, respiration, and atmospheric diffusion, which is affected by wind circulation. Dissolved oxygen levels typically remain high in the surface water because of the availability of light and nutrients that support intense algal growth. However, dissolved oxygen entrapped in the hypolimnic layer usually disappears rapidly through uptake by biological and abiotic processes. The biological oxygen demand (respiration) can be very large in highly productive lakes, causing a rapid depletion of dissolved oxygen in the bottom waters.

Figure 10.7 shows typical temperature and oxygen relationships in a stratified lake throughout the year. During the winter months the lake is effectively isothermal, as is the oxygen gradient. During the summer months a definite thermocline forms, and the lake is effectively partitioned by both temperature and oxygen concentration. The concentration of oxygen is effectively depleted beyond the middle depths of the lake. As the weather cools the upper (epilimnic) waters become cool, the lake once again becomes isothermal (in the autumn), and an overturn occurs.

It is possible to physically destratify a lake using any of several engineering techniques. The main purpose for destratifying a lake is to improve water quality, that is, to upgrade the water quality as a pretreatment to reduce tastes and odors in the water. In addition, artificial aeration can be used to keep northern lakes free of ice, which is sometimes desirable because ice prevents diffusion of oxygen into the lake and kills fish.

There are several approaches to destratifying lakes, and it should be noted that lakes can be either destratified, aerated, or aerated without destratification. In general, artificial destratification is a process by which density differences are decreased between water levels so that prevailing meteorological forces (i.e., winds) can mix the entire lake. On the other hand, it is possible to aerate the hypolimnic waters without destratifying the lake.

There are three basic ways to destratify a lake.

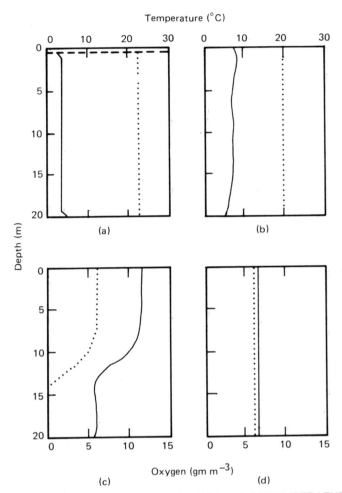

FIGURE 10.7   GENERALIZED SEASONAL PROFILES OF TEMPERATURE AND
OXYGEN IN A TYPICAL STRATIFIED LAKE: (a) January, (b) May, (c) August, (d)
October.   ——, temperature; · · · , oxygen concentration; ---, ice.

(1) *Mechanical pumping.* In this method water from the hypolimnion
is brought to the surface and allowed to settle back to the bottom. Eventually the entire lake becomes mixed.

(2) *Injection of diffused air.* In this case compressed air is introduced to
the bottom of the lake, and as the bubbles rise water is carried along with
them. This results in mixing action and, once again, differences in density
are decreased and the lake becomes totally mixed.

(3) *Aerohydraulics.* This process utilizes an aerohydraulics gun. A low-head, high-volume positive displacement pump allows air bubbles to rise in a vertical tube, promoting a piston action. The rising bubbles force water above it out of the pipe, which then entrains a volume of water behind it in the tube. The result is a narrow turbulent stream that entrains further quantities of water along with it.

In general, all three destratification techniques are required only to disrupt an existing thermocline. Once the lake becomes isothermal, wind currents will generally keep the lake mixed for long periods of time. Although situations are different in different geographical locations, continual energy inputs are usually not necessary to keep a lake destratified. Thus, mechanical equipment is only operated periodically to maintain the lake in an isothermal state.

The biological and chemical changes that occur in a lake during aeration and after destratification are very complex. This is because the normal relationships between the biological and chemical systems in a natural body of water are complex and poorly understood. Perhaps the most basic chemical parameters that influence aqueous systems are alkalinity, carbon dioxide, and pH, and these are usually under biological control. In general, large algal activity (photosynthesis) in the epilimnic water means that free carbon dioxide and some biocarbonate will be depleted, leading to an increase in pH and a decrease in alkalinity. However, in the hypolimnion, where respiration is the principal biological process, there will be an excess of free CO_2, resulting in a decrease in pH and an increase in alkalinity. The relative concentrations of carbon species, as well as pH, will depend on the trophic level of the lake. Highly perturbed lakes (highly eutrophic) show very large differences between epilimnic and hypolimnic CO_2 levels. In general, after aeration or destratification the concentration of free CO_2, total alkalinity, and pH become constant throughout the lake.

The concentrations of the macronutrients nitrogen and phosphorus are also affected when a lake is destratified. Phosphorus tends to redissolve in the anaerobic hypolimnic waters when the lake is stratified, and soluble concentrations tend to be very high. Nitrogen also dissolves in the hypolimnic waters, mostly in the reduced form of ammonium. When a lake is destratified, ammonium nitrogen is circulated throughout the lake and initially decreases in concentration because of the dilution. Biological uptake and nitrification further decreases the ammonium concentration. The concentration of phosphorus also decreases as a lake becomes aerobic. This happens because the principal cations that form complexes with phosphorus, for example, iron and calcium, form insoluble precipitates under aerobic conditions. Thus, phosphate tends to precipitate to the bottom, usually as an iron or calcium salt.

Perhaps one of the most dramatic changes upon the destratification of a lake is the change in algal populations, both in quantity and quality. In general, the total concentration of algae in the epilimnic waters decreases because of the dilution factor. The subsequent production of algae after destratification is variable, and researchers have observed both increases and decreases in total population depending on the situation.

There are many factors that affect algal survival once destratification has occurred. Algae are circulated out of the euphotic zone to the hypolimnic waters, beyond the reach of sunlight, reducing production. However, nutrients from the hypolimnion are introduced to the epilimnion, thus increasing the potential for productivity. The net resultant productivity is therefore difficult to predict and becomes site specific.

One immediate effect of destratification is the elimination of species of algae that depend on the unique temperature stratification of lakes. Other organisms are also affected by destratification, including zooplankton, macroinvertebrates, and fin fish. All of these organisms adapt as much as possible, but new environmental regimes are established upon destratification. Therefore, certain populations of each community are affected. Fin fish are unconfined and can therefore seek a newly established environment favorable to their growth requirements. Invertebrates, especially benthic invertebrates, however, are affected because the previously anaerobic sediments become aerobic after destratification. Most studies have shown that lake sediments generally become more populated with invertebrates after destratification because of the increased oxygen reserve.

The preceeding section described in general the events that occur in a lake after destratification. Several studies have been made of destratified lakes, and many data have been collected that reflect the variability in lake response after destratification. At this point it is not clear whether the benefits from destratification outweigh the side effects of perturbations to the system. It is clear that a highly eutrophic lake receives greater benefits upon destratification than a lake that is less productive. One such comparison of responses to destratification was made by Fast, Moss, and Wetzel (1973) in two Michigan lakes. The two lakes were artificially aerated using compressed air. One test lake was an unproductive lake that was completely mixed, whereas the second lake was highly eutrophic and only the hypolimnion was aerated (i.e., no disruption of the thermocline occurred). Figure 10.8 shows the changes in dissolved oxygen, pH, and alkalinity in the eutrophic lake for the year prior to and the year after hypolimnic aeration. As noted previously, the dissolved oxygen and pH in the bottom waters was

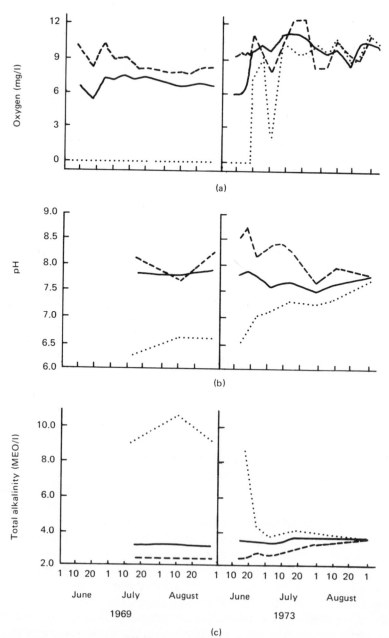

FIGURE 10.8 (A) OXYGEN, (B) PH, AND (C) TOTAL ALKALINITY FOR A EU-
TROPHIC LAKE (HEMLOCK LAKE, MICHIGAN) DURING THE SUMMERS OF
1969 AND 1970.   Continuous artificial aeration was performed between June 14
and September 7, 1970. ---, surface; ——, mean; · · ·, bottom. (From Fast,
Moss, and Wetzel, 1973. © 1973 American Geophysical Union.)

very low before the lake was aerated. After aeration the oxygen concentration in the bottom waters was effectively equivalent to that in the surface waters, so the entire lake was apparently mixed. Also, it is obvious that there was a change in pH and alkalinity, with the resultant values approaching the mean of the bottom and surface waters.

Many chemical and biological parameters were monitored during this study. In general, it was observed that the calcium concentration within the lake declined after aeration, partly owing to dilution but also because of the precipitation of carbonates. The concentration of magnesium after hypolimnic aeration was similar to calcium, and all other cations appeared to be unaffected by aeration.

The total dissolved organic carbon, which was stratified prior to aeration, underwent rapid oxidation after the lake was aerated. The particulate organic carbon, previously concentrated in the hypolimnic waters, was rapidly dispersed after the lake was mixed. There was also some oxidation of the organic fraction after the lake was aerated.

The most important parameter affected by hypolimnic aeration is algal primary productivity. As discussed previously, this is the biological process that tends to regulate most of the chemical and biological features of lakes. Figure 10.9 shows the concentration of phytoplankton, the primary productivity, and the clarity of the water in the lake before and after aeration. Figure 10.9a shows that the water clarity increased after the lake was aerated, but eventually returned to a lower level than that prior to aeration. This response is mirrored by the primary productivity and phytoplankton concentration data. It appears that an immediate clarification of the lake occurred, but that the net result of hypolimnic aeration was a decrease in clarity and an increase in both the primary productivity and the standing crop of biomass. It should be recalled that in this case only the hypolimnic waters were aerated and stratification was supposed to be maintained. However, some leakage occurred in the aeration device, causing some hypolimnic water to reach the epilimnion. This may account for the increase in primary productivity, because the hypolimnic waters contained high concentrations of nutrients.

Figure 10.10 shows the same relationships, that is, pH and alkalinity, for the second lake, which was completely aerated and hence destratified. It should be noted that this lake was in a much lower eutrophic state than the other lake studied. Prior to aeration the oxygen levels were not extremely low, even in the bottom waters. In like manner, both the pH and alkalinity were not extremely different between the epilimnic and hypolimnic waters. However, after aeration the values of these parameters became nearly identical, indicating that the lake was totally destratified.

Data on the transparency and algal productivity of the second lake are

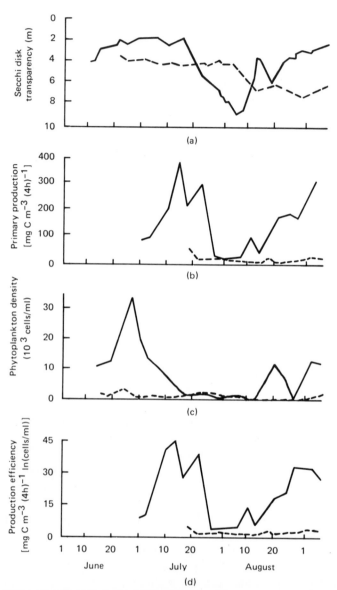

FIGURE 10.9   (a) SECCHI DISK TRANSPARENCY, (b) SURFACE PRIMARY PRODUCTION POTENTIAL, (c) SURFACE PHYTOPLANKTON DENSITY, AND (d) SURFACE PRODUCTION EFFICIENCY DURING THE SUMMERS OF 1969 (---) AND 1970 (——) FOR HEMLOCK LAKE, MICHIGAN.   Continuous artificial aeration was performed between June 14 and September 7, 1970. (From Fast, Moss, and Wetzel, 1973. © 1973 American Geophysical Union.)

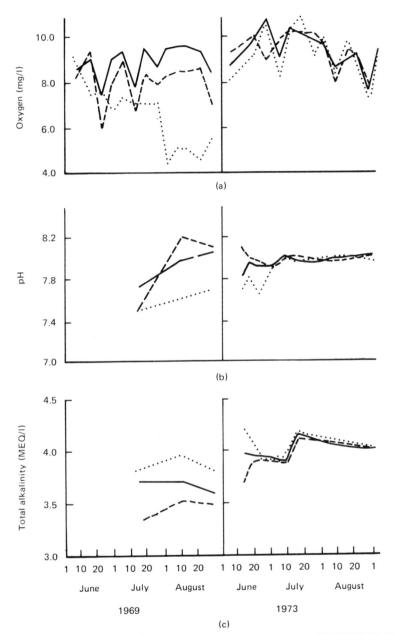

(a)

(b)

(c)

FIGURE 10.10   (a) OXYGEN, (b) PH, AND (c) TOTAL ALKALINITY DURING THE SUMMERS OF 1969 AND 1970 AT SECTION FOUR LAKE, MICHIGAN.   Continuous artificial aeration was performed between June 16 and September 7, 1970. ---, surface; ——, mean; · · ·, bottom. (From Fast, Moss, and Wetzel, 1973. © 1973 American Geophysical Union.)·

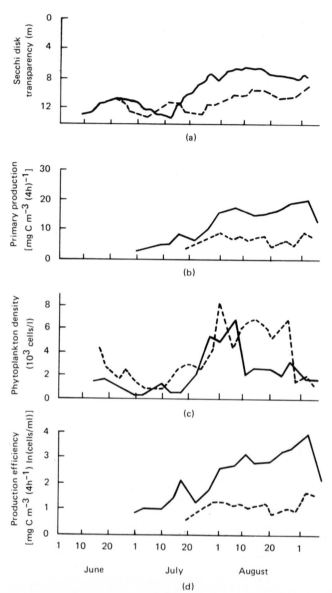

FIGURE 10.11   (a) SECCHI DISK TRANSPARENCY, (b) SURFACE PRIMARY
PRODUCTION POTENTIAL, (c) SURFACE PHYTOPLANKTON DENSITY, AND (d)
SURFACE PRODUCTION EFFICIENCY DURING THE SUMMERS OF 1969 (---)
AND 1970 (——) AT SECTION FOUR LAKE, MICHIGAN.   Continuous artificial
aeration was performed between June 16 and September 7, 1970.   (From Fast,
Moss, and Wetzel, 1973. © 1973 American Geophysical Union.)

shown in Fig. 10.11. Once again, it can be seen that after destratification the clarity decreased and algal production increased. As mentioned previously, this is often the case when nutrient-rich water is introduced into the highly productive euphotic zone. These data are somewhat misleading, however, in that over a period of years the nutrient reservoirs, (i.e., sediments) become effectively capped owing to aerobic conditions in the hypolimnion. The overall productivity is expected to decrease with time, and the clarity to increase with time. In any case, it is apparent that for at least the first year after destratification, the quality of a lake may not be improved.

It can be seen from the previous discussion that the aeration and/or destratification of lakes affects the total lake environment. There may be an improvement in lake quality or the quality may decrease for a short period of time. The basis for continual water quality improvement, of course is that the inputs of nutrients from the sediments to the lake will eventually be precluded, that is, the cycling of both nitrogen and phosphorus will be reduced.

There are other methods of precluding nutrient movement in lakes, one of which is to add to the lake a scavenger chemical that precipitates nutrients. This is most effectively done with phosphorus, because it readily forms insoluble complexes with many cations.

D. *In Situ* Phosphate Precipitation in Lakes

The degradation of lakes due to eutrophication can be slowed by the *in situ* application of nutrient scavengers. Water quality degradation from eutrophic processes is primarily caused by an overgrowth of aquatic plants. In large lakes this is usually observed as excess phytoplankton production. Excess plant growth causes severe diel fluctuations of oxygen, often large fluctuations in alkalinity and pH, and a loss of water clarity. Destratification of a lake as a means of maintaining aerobic sediments to reduce nutrient cycling, however, does not preclude continued plant growth, and mixed results have been observed in aerated lakes.

The principal reason that aquatic weeds grow to excess in lakes is an excess of nutrients. We have already described management approaches for precluding the introduction of nutrients into lakes, but often surface water runoff is sufficient to maintain high nutrient levels. It therefore sometimes becomes necessary to remove these nutrients by the *in situ* application of chemicals. Phosphorus is most often singled out as the nutrient to be removed, because it is often the limiting nutrient in freshwater systems. In addition, phosphorus readily forms insoluble precipitates with certain metals. Nitrogen, on the other hand, seldom forms insoluble complexes and can only be removed from an ecosystem through denitrification or volatilization of ammonia (see Chapter 5). Phosphorus can be precipitated using procedures developed in industry, particularly those used for potable water treatment. The general procedure is to add a trivalent metallic ion, such as aluminum or iron, to form an aluminum or iron phosphate complex.

In one study Peterson *et al.* (1973) treated a eutrophic lake in Wisconsin with aluminum sulfate. The lake behaved as a typical temperate lake, with thermoclines forming in the summer months and having an ice cover in the winter. The lake exhibited all of the common water quality characteristics that classify lakes as highly eutrophic and in need of some type of water quality restoration. Table 10.12 shows the physical and chemical data for the lake prior to treatment for phosphorus removal. Note that there are distinct differences in water quality between the epilimnic and hypolimnic waters, described previously. In particular, it is apparent that high concentrations of nitrogen and phosphorus remain dissolved in the hypolimnic waters, which is a result of the very low dissolved oxygen concentrations.

TABLE 10.12   RANGE OF PHYSIOCHEMICAL
DATA FOR HORSESHOE LAKE[a]

PARAMETER	EPILIMNION	HYPOLIMNION
Temperature in °F	32–78	32–48
Dissolved oxygen (mg l^{-1})	<0.5–14.3[b]	0–4.3
pH	7.2–8.9	6.8–8.3
Total alkalinity (mg CaCO$_3$ l^{-1})	218–252	220–278
Total hardness (mg CaCO$_3$ l^{-1})	254–300	276–306
Nitrite N (mg l^{-1})	0.004–0.089	0.002–0.213
Nitrate N (mg l^{-1})	0.10–0.90	0.10–0.76
Ammonia N (mg l^{-1})	0.06–1.96	0.16–6.33
Organic N (mg l^{-1})	0.76–2.19	0.75–1.68
Dissolved P (mg l^{-1})	0.01–0.48	0.26–1.52
Total P (mg l^{-1})	0.05–0.50	0.31–1.54

[a] From Peterson *et al.* (1973). The data were taken from files of the Wisconsin Department of Natural Resources, State Laboratory of Hygiene. The data were gathered from January, 1966 to April, 1967.
[b] The minimum oxygen value was taken from field data on file with the Wisconsin Department of Natural Resources, Area Office, Plymouth (1965).

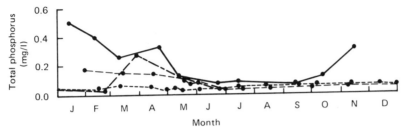

FIGURE 10.12 EPILIMNIC TOTAL PHOSPHORUS CONCENTRATION FOR HORSESHOE LAKE, WISCONSIN, BEFORE AND AFTER TREATMENT WITH ALUMINUM SULFATE TO REMOVE PHOSPHORUS ON MAY 20, 1970. ——, 1966; ———, 1970; ---, 1971; ---, 1972. (From Peterson *et al.*, 1973.)

The treatment equipment consisted of tanks for mixing aluminum sulfate slurries, a supply of fresh water for feeding these tanks, a mixer, and a manifold supported on the back of small boats, placed approximately 1 ft below the water surface. The treatment concentration of aluminum sulfate was determined previously in laboratory experiments, and that concentration was applied to the entire lake using the specially equipped boats. After the aluminum sulfate was applied, the water quality of the lake and several biological parameters were monitored for 2 years. Figures 10.12 and 10.13 show total phosphorus concentrations in the epilimnic and hypolimnic waters of the lake before and after treatment. There was little apparent change in the epilimnic waters other than in residual phosphorus, which had a slightly lower concentration during the early spring months. In the hypolimnic waters, however, residual phosphorus concentrations were reduced by the additiɔn of aluminum sulfate and remained at a reduced level even 2 years after treatment. It apears that the aluminum precipitate that ultimately ended up on the bottom of the lake formed some kind of cap on the sediments, precluding the release of phosphorus. Data from this experiment showed that approximately 10 to 15 lb of phosphorus were immediately precipitated from the lake upon the addition of the aluminum salts. Using the average for the entire lake, the total phosphorus removed from the lake amounted to approximately 110 lb after 2 years. An economic analysis of the treatment showed that unless more phosphorus were removed, it would probably not be economically feasible to use the method to improve water quality.

The most striking result of the addition of aluminum sulfate to the Wisconsin lake was an increase in the dissolved oxygen concentration in the hypolimnic water. Prior to treatment the dissolved oxygen was essentially depleted throughout the lake during each winter when ice was formed.

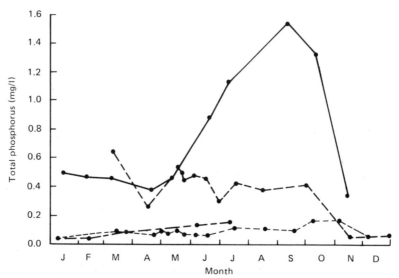

FIGURE 10.13 HYPOLIMNIC TOTAL PHOSPHORUS CONCENTRATION FOR HORSESHOE LAKE, WISCONSIN, BEFORE AND AFTER TREATMENT WITH ALUMINUM SULFATE TO REMOVE PHOSPHORUS ON MAY 20, 1970. ——, 1966; –––, 1970; ---, 1971; –––, 1972. (From Peterson et al., 1973.)

After treatment, however, there was sufficient oxygen in the entire lake to maintain a sport fishery.

The nitrogen concentration in the lake did not vary appreciably between pretreatment and postreatment samples, but there is little reason to expect nitrogen concentration to change. The conclusions of this study were as follows.

(1) A decrease in the total phosphorus content of the lake occurred during the summers following treatment.

(2) There was no large increase in the total phosphorus content of the hypolimnion during the following two summers.

(3) There was some increase in the transparency of the water.

(4) There was a short-term decrease in the color of the water.

(5) There was an absence of the nuisance algal blooms that had been common in the lake prior to treatment.

(6) There was a marked improvement in the concentration of dissolved oxygen, especially during the winter.

(7) There were no observed adverse ecological affects from the chemical addition.

The study was carried out on a temperate lake that suffered eutrophication primarily because of excess phytoplankton growth. Many shallow lakes show a similar degradation in water quality because of large growths of benthic or rooted plants. One such series of lakes was analyzed in Florida by Haumann and Waite (1978), and phosphorus was once again removed from the lakes by the addition of aluminum sulfate. Simultaneous analyses of plant production, physiochemical removal, and hydraulic flushing were performed, and Fig. 10.14 summarizes the removal mechanisms in the lakes. There was a very rapid removal of phosphorus upon addition of the aluminum sulfate as the material precipitated to the bottom. After approximately 11 days the total removal began to proceed at a slower rate. It was shown that the majority of phosphorus removal in the initial stages of the

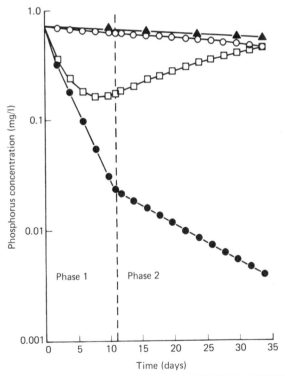

FIGURE 10.14   BREAKDOWN OF PHOSPHATE REMOVAL USING ALUMINUM SULFATE FOR A SERIES OF SHALLOW FLORIDA LAKES. o, total removal; △, physical and chemical removal; o, biogenic removal; □, hydraulic flushing. (From Haumann and Waite, 1978.)

experiment were indeed physiochemical in nature, that is, phosphorus was precipitated by aluminum.   After a period of approximately 2 weeks, however, the rate of phosphorus removal by benthic plants exceeded the rate of continued physiochemical removal.   Analysis of the sediments during this period showed that the majority of phosphorus and aluminum did end up in the sediments, and it is assumed that this phosphorus became available to the rooted benthic plants.   However, it appeared that the benthic community was not able to use the phosphorus from the complex aluminum phosphate sediment as readily as it could from the phosphorus in the water column.   It was therefore concluded that adding aluminum to the water to precipitate phosphorus could reduce the biomass potential of a lake. However, the total reduction was not determined during this experiment.

REFERENCES

Baker, M. N. (1949).   "The Quest for Pure Water."   The American Water Works Association, New York.

Bryan, E. H. (1970).   Quality of stormwater drainage from urban land areas in North Carolina.   Rpt. No. 37., Water Resources Research Institute, University of North Carolina, Chapel Hill, North Carolina.

Cabelli, V. J., and McCabe, L. J. (1974).   "Recreational Water Quality Criteria."   Environmental Protection Agency, Washington, D.C.

Craun, G. F. (1977).   Waterborne outbreaks.   *J. Water Pollut. Control. Fed.* **49,** 1268.

Craun, G. F., and McCabe, L. J. (1973).   Review of causes of waterborne disease outbreaks. *J. Am. Water Works Assoc.* **65,** 74.

Craun, G. F., *et al.* (1976).   Waterborne disease outbreaks in the U.S.   *J. Am. Water Works Assoc.* **68,** 420.

Edmondson, W. T. (1970).   Phosphorus, nitrogen and algae in Lake Washington after diversion of sewage. *Science* **169,** 690.

Emery, R. M., Moon, C. E., and Welch, E. B. (1973).   Delayed recovery of a mesotrophic lake ' after nutrient diversion. *J. Water Pollut. Control Fed.* **45**(5).

Environmental Protection Agency (1970).   "Storm Water Pollution from Urban Land Activity," *Water* Pollution Control Research Series, 11034 FKL 07/70.   Environmental Protection Agency, Washington, D.C.

Fast, H. W., Moss, B., and Wetzel, R. G. (1973).   Effects of artificial aeration on the chemistry and algae of two Michigan lakes. *Water Res. Res.* **9**(3).

Geldreich, E. E. (1970).   Applying bacteriological parameters to recreational water quality. *J. Am. Water Works Assoc.* **62**(2).

Haumann, D., and Waite, T. D. (1978).   The kinetics of phosphate removal in small alkaline lakes by natural and artificial processes. *Water, Air, Soil Pollut.* **10**.

Hollis, D. G., Weaver, R. E., Baker, C. N., and Thornesberry, C. (1976).   Halophilic vibrio species isolated from blood cultures. *J. Clin. Microbiol.* **3,** 425.

Lee, G. F. (1966).   "Report on the nutrient sources of Lake Mendota."   Lake Mendota Problems Committee, Madison, Wisconsin.

McGauhey, P. H. (1968). "Engineering Management of Water Quality." McGraw-Hill, New York.

McKee, J. E., and Wolf, H. W. (1963). "Water Quality Criteria." Publication 3-A, The California State Water Quality Control Board, Sacramento, California.

Peterson, J. O., Wall, J. B., Wirth, T. L., and Born, S. M. (1973). Nutrient inactivation by chemical precipitation at Horseshoe Lake, Wisconsin. Technical Bulletin No. 62, Department of Natural Resources, Madison, Wisconsin.

Römkens, M. J. M., and Nelson, D. W. (1974). Phosphorus relationships in runoff from fertilized soils. *J. Environ. Qual.* 3(1).

Stokes, W. R. (1927). Report of the committee on bathing places. *Am. J. Public Health* 17, 334.

Wallis, I. G. (1973). Options for improving water quality. *Int. J. Environ. Stud.* 6.

Pipes, W. O., ed. (1978). "Water Quality and Health Significance of Bacterial Indicators of Pollution." Proceedings of a National Science Foundation Workshop held at Drexel University, Philadelphia, Pennsylvania.

Index

285